# Plants of Central Asia
## Volume 13

# Plants of Central Asia

*Plant Collections from China and Mongolia*

(*Editor-in-Chief:* V.I. Grubov)

## Volume 13

### Plumbaginaceae, Oleaceae, Buddlejaceae, Gentianaceae, Menyanthaceae, Apocynaceae, Asclepiadaceae

V.I. Grubov

CRC Press
Taylor & Francis Group
Boca Raton London New York

CRC Press is an imprint of the
Taylor & Francis Group, an **informa** business

A SCIENCE PUBLISHERS BOOK

First published 2007 by Science Publishers Inc.

Published 2019 by CRC Press
Taylor & Francis Group
6000 Broken Sound Parkway NW, Suite 300
Boca Raton, FL 33487-2742

© 2007, Copyright Reserved
CRC Press is an imprint of Taylor & Francis Group, an Informa business

First issued in paperback 2019

No claim to original U.S. Government works

ISBN 13: 978-0-367-45321-3 (pbk)
ISBN 13: 978-1-57808-421-0 (hbk)
ISBN 13: 978-1-57808-062-5 (Set)

**Visit the Taylor & Francis Web site at**
http://www.taylorandfrancis.com

**and the CRC Press Web site at**
http://www.crcpress.com

Library of Congress Cataloging-in-Publication Data
Rasteniia Tsentral'noĭ Azii. English
  Plants of Central Asia: plant collections from
China and Mongolia
/[editor-in-chief, V.I. Grubov].
    p. cm.
  Research based on the collections of the V.L. Komarov
Botanical Institute.
  Includes bibliographical references.
  Contents: V.13. Plumbaginaceae, Oleaceae, Buddlejaceae,
Gentianaceae, Menyanthaceae, Apocynaceae, Asclepiadaceae
  ISBN 978-1-57808-421-0 (vol. 13)
  1. Botany-Asia, Central. I. Grubov, V.I. II.
Botanicheskiĭ institut im. V.L. Komarova. III. Title.
QK374, R23613 2002
581.958-dc21
                                        99-36729
                                        CIP

ACADEMIA SCIENTIARUM ROSSICA
INSTITUTUM BOTANICUM nomine V.L. KOMAROVII
PLANTAE ASIAE CENTRALIS
(secus materies Instituti botanici nomine V.L. Komarovii)
Fasciculus 13
PLUMBAGINACEAE, OLEACEAE, BUDDLEJACEAE, GENTIANACEAE,
MENYANTHACEAE, APOCYNACEAE, ASCLEPIADACEAE
Confecerunt: V.I. Grubov

Translation of:  Rasteniya Tsentral'noi Azii, vol. 13. 2002
                 Izdatel'stvo Sankt-Petersburgskoi gosundarstvennoi
                 khimiko-farmatsevticheskoi akademii (St. Petersburg
                 State Chemical-Pharmaceutical Academy Press),
                 St. Petersburg

# ANNOTATION

PLANTS OF CENTRAL ASIA. From the Material of the V.L. Komarov Botanical Institute, Russian Academy of Sciences, vol. 13: Plumbaginaceae, Oleaceae, Buddlejaceae, Gentianaceae, Menyanthaceae, Apocynaceae and Asclepiadaceae. Compiler: V.I. Grubov. 2002. Izdatel'stvo sankt-peterburgskoi gosudarstvennoi khimiko-farmatsevticheskoi akademii (St.-Petersburg State Chemical-Pharmaceutical Academy Press).

This volume of the illustrated lists of flowering plants of Central Asia (within the People's Republics of China and Mongolia) treats leadwort (Plumbaginaceae), olive (Oleaceae), butterfly-bush (Buddlejaceae), gentian (Gentianaceae), buck-bean (Menyanthaceae), dogbane (Apocynaceae) and milkweed (Asclepiadaceae) families. The more abundant among these, leadworts (41 species) and gentians (111 species), play a significant role in the vegetative cover: the former in the desert territories and the latter in the high mountains. They are also of interest in the botanical-geographic context.

As in the preceding volumes, keys are provided for the identification of genera and species. References of nomenclature and information on ecology and geographic distribution are given for each species.

Ill.: 4 plates, 5 maps of distribution ranges.

V.I. Grubov
Editor-in-Chief

# PREFACE

This volume covers the treatment of 7 families of gamopetalous flowering plants: Plumbaginaceae Juss. (7 genera, 41 species), Oleaceae Hoffmgg. (3, 6), Buddlejaceae Wilh. (1, 5), Gentianaceae Juss. (8, 111), Menyanthaceae Dum. (2, 2), Apocynaceae Juss. (1, 2) and Asclepiadaceae R. Br. (2, 14). In all, these families comprise 24 genera and 180 species, with 34 of them being endemics and 30 subendemics. To these should be added one more genus (*Psylliostachys* Nevski) with 21 species (including 6 endemics) from the Central Asian part of the erstwhile USSR, thus bringing the total number of species inhabiting the entire Central Asia will be 201; 40 of these, or 20%, are endemics.

Of the families treated in this volume, the largest and the important in the flora are gentians (Gentianaceae) and leadworts (Plumbaginaceae). These 2 families represent antipodes in the ecological and coenotic context: while gentians are mesophytes and predominantly inhabit alpine meadows, leadworts are desert and arid-steppe xero- and halophytes. Among gentians, only *Gentiana dahurica, G. decumbens* and *G. olivieri* can be termed as steppe inhabitants. Gentians in Central Asia are representatives and derivatives of East Asian flora, only 10 of them being endemics (11%). Quite naturally, maximum diversity among gentians is noticed on the eastern fringe of Central Asia—in Qinghai and northeastern Tibet (Weitzan) as well as in Tien Shan. All of the endemics are concentrated here.

Leadworts, however, represent Mediterranean flora and are rich in local endemics: of the 41 species of leadworts, 18 or 44% are endemics and, on adding to them the 11 subendemic species, endemism rises to 70% ! Genus *Limonium* Mill. contains the maximum number of endemic species: of the 23 species, 13 are endemics and 5 subendemics.

The coenotic importance of both these families in the vegetative cover of Central Asia is not of much significance. Many gentians are highly ornamental plants and add colour to the alpine meadows while a few of them sometimes make for a colourful scenario, as for example *Gentiana barbata* in the wet coastal meadows.

Massive growth of some species of leadworts of genus *Limonium* make up their own scenario: *L. bicolor* (pink inflorescence) in the steppes of Eastern Mongolia and *L. tenellum* (blue) and *L. aureum* (lemon-yellow) in

4

desert steppes and deserts of Mongolia. Species of *Acantholimon* Boiss. occupy a prominent position in the pillow structures of Tien Shan, Pamir and western Tibet.

Representatives of rest of 5 families are few, rare and very limited in their spread. Only dogbane (*Apocynum* L.) and some swallowworts (*Cynanchum* L. of Asclepiadaceae) are extensively distributed. *A. venatum* and *A. pictum*, inhabitants of solonchaks, are found in all desert regions from Junggar and Kashgar to Alashan and Qinghai. Liana *Cynanchum sibiricum* is widely distributed in the oases of Central Asia while *C. thesioides* (= *Vincetoxicum sibiricum*) is a common sight in desert and arid steppes of Mongolia and southern Siberia. The rest of members of this large family represent only rare inhabitants of East Asia, mainly in the border regions: some of them like *Periploca sepium* and *Cynanchum bungei* are distinct relics while others, for example *C. chinense*, are confined to their own distribution range. *C. gobicum*, *C. komarovii* and *C. pusillum* have, however, succeeded in developing into local endemics here.

The family of olives is represented in Central Asia by 3 Chinese species of lilac in Alashan mountain range and mountains of Qinghai and are of relict type: Mediterranean jasmine (*Jasminum officinale*) and 2 species of Mediterranean ash, *Fraxinus sogdiana* in tugais of Junggar and Kashgar and Himalayan *F. xanthoxyloides* found in extreme south-western Tibet.

The large (over a hundred species) East Asian, predominantly tropical genus *Buddleja* L. is represented by the lone Chinese species *B. alternifolia*, a relict of Alashan mountain range and Eastern Qinghai and 4 species in south-eastern Tibet in Lhasa region, having reached there from East-Tibet and eastern Himalayas. One of them is a local derivative-endemic (*B. wardii*). *Menyanthes trifoliata* and *Nymphoides peltatum* represent boreal, marsh and coastal-aquatic, extensively distributed species.

Among the chronologically interesting data is the distribution of Tibet-Qinghai species of gentians: annual *Gentiana arenaria* Maxim., reported quite recently from Central Tien Shan (Sarydzhas river basin) and *G. saposhnikovii* Pachom., detected and described as a new species (1986). It is possible that, the latter, like *Circaeaster agrestis* Maxim. from among buttercups, will be found in the Himalayas too (the plant is small and escapes detection) which would once again confirm the historic links between Tien Shan and Tibet through the Himalayas.

The distribution of leadworts is of interest. Several species of leadworts form well-distinguished, distinct botanical-geographic limits of different ranks. Thus, for example, *Limonium tenellium* (Turcz.) Ktze. is an endemic of Mongolia with its distribution covering its entire Gobi territory while *L. roborowskii* Ik.-Gal. is an endemic of Kashgar with distribution coinciding with its boundaries. The distribution of *L. myrianthum* (Schrenk) Ktze. encompasses the whole of Junggar-North Turan province from Aralo-Caspian region in the west of Junggar Gobi in the east; that of *L. aureum* (L.) Hill covers the entire Mongolian-Chinese Central Asia, barring Sinkiang. *Acantholimon alatavicum* Bge. marks the mountainous Junggar and *A. kokandense* Bge. only Western Kashgar.

5

In this volume, O.I. Starikova translated the Chinese references and herbarium labels. Artist O.V. Zaitseva prepared the drawings for plates. I.D. Illarionova plotted the maps of distribution ranges.

# CONTENTS

# TAXONOMY

## SPECIAL ABBREVIATIONS

### Abbreviations of Names of Collectors

| | |
|---|---|
| Bar. | — V.I. Baranov |
| Chaff. | — J. Chaffanjon |
| Chaney | — R.W. Chaney |
| Chen and Chou | — Chen Tszin-shen and Chou Guan-yui |
| Ching | — R.C. Ching |
| Chu | — C.N. Chu |
| Czern. | — E.G. Czerniakovskaya |
| Czet. | — S.S. Czetyrkin |
| Divn. | — D.A. Divnogorskaya |
| Fet. | — A.M. Fetisov (Fetisow) |
| Glag. | — S.A. Glagolev |
| Grub. | — V.I. Grubov |
| Gr.-Grzh. | — G.E. Grum-Grzhimailo |
| Gub. | — I.A. Gubanov |
| Gus. | — V.A. Gusev |
| Hao | — Hao Kin-Shen |
| Ik.-Gal. | — N.P. and V.A. Ikonnikov-Galitzkij |
| Isach. | — E.A. Isachenko (also known as E.A. Volkova) |
| Ivan. | — A.F. Ivanov |
| Kal. | — A.V. Kalinina |
| Kam. | — R.V. Kamelin |
| Karam. | — Z.V. Karamysheva |
| Klem. | — E.N. Klements |
| Kom. | — V.L. Komarov |
| Krasch. | — I.M. Krascheninnikov |
| Kryl. | — P.N. Krylov |

| | | |
|---|---|---|
| Kuan | — | K.C. Kuan |
| Lad. | — | V.F. Ladygin |
| Ladyzh. | — | M.V. Ladyzhevsky |
| Lavr. | — | E.M. Lavrenko |
| Lee and Chu, Li et al. | — | A.R. Lee (1958) |
| Lee | — | S.-y. Lee |
| Liou | — | Liou Tchen-ngo |
| Lis. | — | V.I. Lisovsky |
| Litw. | — | D.N. Litwinow |
| Martin | — | J. Martin |
| Merzb. | — | G. Merzbacher |
| Mois. | — | V.S. Moiseenko |
| Muld. | — | A.A. Muldashev |
| Pal. | — | I.V. Palibin |
| Pavl. | — | N.V. Pavlov |
| Petr. | — | M.P. Petrov |
| Pewz. | — | M.V. Pewzov |
| Pias. | — | P.Ya. Piassezki |
| Pob. | — | E.G. Pobedimova |
| Pop. | — | M.G. Popov |
| Pot. | — | G.N. Potanin |
| Przew. | — | N.M. Przewalsky (Przewalskyi) |
| Rachk. | — | E.I. Rachkovskaya |
| Reg. | — | A. Regel |
| Rob. | — | V.I. Roborowsky |
| Sap. | — | V.V. Sapozhnikov |
| Schischk. | — | B.K. Schischkin |
| Serp. | — | V.I. Serpukhov |
| Shen | — | Sheng Tian |
| Sold. | — | V.V. Soldatov |
| Tug. | — | A.Ya. Tugarinov |
| Ulzij. | — | N. Ulzijkhutag |
| Yuan' | — | Yi-f. Yuan |
| Yun. | — | A.A. Yunatov |
| Zab. | — | D.K. Zabolotnyi |
| Zam. | — | B.M. Zamatkinov |

# Abbreviations of the Names of Herbaria

| | | |
|---|---|---|
| B | — | Botanisches Museum, Berlin-Dahlem |
| BM | — | British Museum of Natural History, London |
| C | — | Botanical Museum and Herbarium, Copenhagen |
| Cal. | — | Central National Herbarium, Botanical Survey of India, Calcutta (Kolkata) |
| CGE | — | Botany School, University of Cambridge |
| E | — | Royal Botanic Garden, Edinburgh |
| G | — | Conservatoire et Jardin botaniques, Geneve (Geneva) |
| GOET | — | Systematisch-Geobotanisches Institut, Universitat Göttingen |
| HNWP | — | Herbarium, Northwest Plateau Institute of Biology, Academia Sinica, Xining, Qinghai |
| K | — | The Herbarium, Royal Botanic Gardens, Kew, London |
| Kun. | — | Kunming Station of the Botanical Institute, Academia Sinica, Kunming |
| Linn. | — | Herbarium, The Linnean Society of London, London |
| Liv. | — | City of Liverpool Museum, Liverpool |
| M | — | Botanische Staatsammlung, München |
| MW | — | Herbarium of the Moscow State University, Moscow |
| P | — | Museum National d'Histoire Naturelle, Laboratoire de Phanerogamie, Paris |
| PE | — | Herbarium, Institute of Botany, Academia Sinica, Beijing |
| PR | — | Botanical Department of National Museum, Praha |
| S | — | Section for Botany, Swedish Museum of Natural History, Stockholm |
| SZ | — | Herbarium of the Department of Biology, National Szechuan University, Chongtu |
| W | — | Naturhistorisches Museum, Botanische Abteilung, Wien (Vienna) |

## Family 93. Plumbaginaceae Juss.

1. Style adnate with 5 long stigmas, bearing large capitate glands inside. Calyx herbaceous, green, without limb ...................................... PLUMBAGOIDEAE ................................. 2.

+ Style free from base; stigma short, fine-glandular all over surface. Calyx coriaceous or membranous, bright-coloured ........ ...................................... STATICOIDEAE ................................... 3.

4

2. Calyx with large stalked glandules in upper half; inflorescence spicate; flowers small, up to 6 mm long. Annuals ..............
.............. 1. Plumbagella Spach (*P. micrantha* (Ldb.) Spach).

+ Calyx without glandules; inflorescence short-spicate to capitate. Shrubs ...................................................... 2. Ceratostigma Bge.

3. Stigma hemispherical-capitate or elongated-capitate .............. 4.

+ Stigma filiform. Plant with radical leaves .................. 6.

8  4. Cushion-forming or creeping subshrubs, mostly with spiny rigid cauline leaves and almost always with many peduncles. Ovary fusiform or narrow-ovoid, gradually narrowed into style ...................................................................... 5.

+ Herbaceous perennials with radical rosette of fairly broad, non-spiny leaves and solitary or very few (2–3, very rarely up to 10) peduncles. Ovary ovoid, slightly narrowed upward, bearing villous styles .................... 5. Goniolimon Boiss.

5. Leaves linear-lanceolate to oblong-obovoid, 3–8 (10) cm long, 0.6–1.5 (2.5) cm broad, prominently crispate along margin, with short cusp; styles verruculose in lower half. Plant 15–20 cm tall, with short procumbent stalks and strong peduncles .................
.................. 4. Ikonnikovia Lincz. (*I. kaufmanniana* (Rgl.) Lincz.).

+ Leaves linear to acicular and subulate, close-sessile; styles glabrous. Cushion-forming or creeping, 3–15 cm tall subshrubs .................................................... 3. Acantholimon Boiss.

6. Flowers on compact, globose individual heads, terminating in simple, erect, leafless, fairly tall peduncles; leaves narrow-linear .................. 6. Armeria Willd. (*A. sibirica* Turcz.).

+ Flowers on spikelets with one or more flowers, usually spicate (rarely subcapitate) on branches of paniculate or corymbose-paniculate inflorescences; often with barren lower twigs on branched peduncles; leaves linear-spatulate and lanceolate to broad-obovoid and obdeltoid .................. 7. Limonium Mill.

## 1. Plumbagella Spach
Hist. natur. veg. phan. 10 (1841) 333, in nota

1. P. micrantha (Ldb.) Spach, Hist. natur. veg. phan. 10 (1841) 333; Boiss. in DC. Prodr. 12 (1848) 690; Kryl. Fl. Zap. Sib. 9 (1937) 2152; Walker in Contribs U.S. Nat. Herb. 28 (1941) 650; Shteinb. in Fl. SSSR, 18 (1952) 297; Grub. Konsp. fl. MNR [Conspectus of Flora of Mongolian People's Republic] (1955) 221; Fl. Kazakhst. 7 (1964) 48; Grub. Opred. rast. Mong. [Key to Plants of Mongolia] (1982) 198; Fl. Xizang, 3 (1986) 847; Peng in Fl. Sin. 60, 1 (1987) 9; Lincz. in Opred. rast. Sr. Azii [Key to

Plants of Mid. Asia] 10 (1993) 6. — *Plumbago micrantha* Ldb. Fl. alt. 1 (1829) 171; ej. Fl. Ross. 3 (1849) 471. — *P. spinosa* Hao in Feddes Repert. 36 (1934) 222 and in Contribs Inst. Bot. Nat. Ac. Peiping, 3 (1935) 2; id. in Bot. Jahrb. 68 (1938) 627.

— Ic.: Ldb. Ic. pl. fl. ross. 1, tab. 21; Fl. Xizang, 3, tab. 328, fig. 1–3; Fl. Sin. 60, 1, tab. 1, fig. 7–8.

Described from West. Siberia (Altay). Type in St.-Petersburg (LE).

On rocky slopes, talus and rocks, more often as ruderal and weed in plantations, fallow lands and abandoned cultivated fields, around residences, roadsides, up to 3800 m.

IA. Mongolia: *Khesi* (Loukhu-shan' mountain range, among hemp plantations, July 17, 1908 — Czet.).

IC. Qaidam: *Mount.* (Dulan-khit temple, 3350 m, in open glades in spruce forests, Aug. 8-9, 1901 — Lad.).

IIA. Junggar: *Tien Shan* (facing Aryslyn estuary, 2450 m, July 8; left bank of Kash river, 2750 m, July 15–1879, A. Reg.).

IIIA. Qinghai: *Nanshan* (lower forest belt of mountain range south of Tetung river, 2300 m, Aug. 7; lower forest belt of mountain range north of Tetung river, 2450-2750 m, Aug. 18–1880, Przew.; "Lien-Cheng, on grassland, No. 312, July 1923, R.C. Ching" — Walker, l.c.), *Amdo* ("Kokonor, Dahoba, zwischen Felsen um 4000 m, No. 1180, Sept. 7, 1930" — Hao, l.c., 1935).

IIIB. Tibet: *Weitzan* (Yangtze river basin, Nru-chyu area, 3550 m, July 25; Yangtze river basin, in I-chyu river valley, 3800 m, July 28–1900, Lad.).

General distribution: Jung.-Tarb., Nor Tien Shan; West. Sib. (Altay), East. Sib. (Daur.), Nor. Mong. (Hang., Mong. Daur.), China (Nor.-West.: Gansu, South-West., nor.-west. Sichuan).

## 2. Ceratostigma Bge.

Enum. pl. China bor. (1832) 55; id. in Mem. Ac. Sci. St.-Petersb. Sav. Etrang. 2 (1835) 129; Lawrence in Gentes Herb. 8 (1954) 410–420. — *Valoradia* Hochst. in Flora, 25 (1842) 239.

1. Spiny, up to 20 cm or more tall, subshrubs; bud scales modified into subulate or acicular stiff spines .............. 3. C. ulicina Prain.

+ Non-spiny, 0.3–1.5 m tall, shrubs; bud scales ovoid, soft, small ......................................................................................... 2.

2. Evergreen shrub. Leaves compactly covered on both sides with fascicular hairs. Young shoots brown ...... 1. C. griffithii Clarke.

+ Deciduous shrub. Leaves glabrous or with diffusely pubescent simple hairs. Young shoots sometimes compactly grey-haired ................................................................. 2. C. minus Stapf.

1. C. griffithii Clarke in Hook. f. Fl. Brit. India, 3 (1882) 481; Prain in J. Bot. Brit. a. For. 44 (1906) 7; Fl. Xizang, 3 (1986) 850; Peng in Fl. Sin. 60, 1 (1987) 11.

—Ic.: Fl. Xizang, 3, tab. 323, fig. 10–11; Fl. Sin. 60, 1, tab. 2, fig. 1–3.

Described from Himalayas (Bhutan). Type in London (K).

Along gorges and among scrubs on mountain slopes up to 4000 m.

IIIB. Tibet: *South.* ("Shigatsze" — Fl. Xizang, l.c.).

General distribution: China (South-West. — south-west. Sichuan, nor.-west. Yunnan), Himalayas (east.).

2. C. minus Stapf ex Prain in J. Bot. Brit. a. For. 44 (1906) 7, p.p. excl. pl. e Tibet occ.; Marq. in J. Linn. Soc. Bot. (London) 48 (1929) 200; Hand.-Mazz. Symb. Sin. 7 (1936) 731; Svenson in Brittonia, 8, 1 (1954) 56; Fl. Xizang, 3 (1986) 849; Peng in Fl. Sin. 60, 1 (1987) 12.

—Ic.: Fl. Sin. 60, 1, tab. 2, fig. 4–7.

Described from South-West. China (vicinity of Kandin). Type in London (BM ?).

In arid gorges and sandy-rubbly mountain slopes, 2600–4000 m.

IIIB. Tibet: *South.* ("Lhasa" — Fl. Xizang, l.c.).

General distribution: China (South-West. — west. Sichuan, cent. and south. Yunnan, Kam).

3. C. ulicina Prain in J. Bot. Brit. a. For. 44 (1906) 7; Fl. Xizang, 3, (1986) 849; Peng in Fl. Sin. 60, 1 (1987) 10.

—Ic.: Fl. Xizang, 3, tab. 328, fig. 8–9; Fl. Sin. 60, 1, tab. 2, fig. 12.

Described from South. Tibet (vicinity of Tszyantszy town). Type in London (BM ?).

Along sandy-rubbly mountain slopes and sandy-pebbly banks of rivers and lakes, 3800–4500 m.

IIIB. Tibet: *South.* ("Shigatsze, Nan'mulin, Nimu" — Fl. Xizang, l.c.; "Tszyantszy" — Fl. Sin. l.c.).

General distribution: China (South.-West. — Kam).

## 3. Acantholimon Boiss.
### Diagn. pl. or., ser. 1, 7 (1846) 69

1.  "Cushions" compact. Leaves compactly sessile, short-linear, terete, obtuse or short-cuspidate, up to 10 mm long. Peduncles either not developed at all (and spikes sessile) or only slightly longer than leaves and barely rise above "cushion". All spikelets 2–4-flowered and with more than 3 bracts; only sometimes, upper spikelets in inflorescence 1-flowered and with 3 bracts ............................................................................. 2.

+  "Cushions" loose. Leaves spaced, linear, acicular or linear-lanceolate, acute spinelike, 1.5–4.5 cm long. Peduncles 3–8, up to 15, cm tall, rise prominently above "cushion" ...................... 4.

2. Peduncles not developed and spikes or some spikelets sessile. Spikelets 1–2-flowered. Calyx tube rather sparsely pubescent only along nerves. Leaves very small, 1–8 mm long, terete-trigonous ............................................................................................ 3.

+ Peduncles, although short, distinctly manifest; only slightly longer than leaves, up to 1.5 cm tall. Spikelets with 2 or more flowers. Calyx tube compactly pubescent, its limb white ...........
................................................................. 2. A. borodinii Krassn.

3. Leaves 1–3 (4) mm long, 0.5–1 mm broad, fleshy-trigonous, subobtuse, without cusp. Flowers in single sessile spikelets or in inflorescences with 2–3 spikelets; calyx 5–6 mm long, with white limb .............................................. 3. A. diapensioides Boiss.

+ Leaves 4–8 mm long, 1–1.5 mm broad, with very short cusp or subobtuse. Flowers in inflorescences with 2–3 spikelets or on single spikelets; calyx 7–8 mm long, its limb dark raspberry to pink and white with raspberry-red nerves ..................................
.................................................................. 4. A. hedinii Ostenf.

4. All spikelets 1-flowered, with 3 bracts ..................................... 5.

+ All spikelets 2–4-flowered, with more than 3 bracts ................ 6.

5. Leaves glabrous, usually ciliate-scabrous only along margin, light green or glaucescent, about 1 mm broad. Peduncles simple, only slightly longer than leaves, 3–6 cm tall. Limb of calyx white, corolla bright pink ................ 1. A. alatavicum Bge.

+ Leaves short and more or less compactly pubescent, bright green, very thin, about 0.5 mm broad. Peduncles generally branched at tip, with compact double row of spikes up to 2–4 cm long, considerably longer than leaves, 4–8, up to 15 cm tall. Limb of calyx white or sometimes with narrow band only along nerve, corolla pink ..................................... 5. A. kokandense Bge.

6. Leaves linear-lanceolate, 1.5–3 mm broad, flat, with cusp, highly deflexed. Peduncles 3–5 cm tall, with compact spike bearing 5–8 spikelets in 2 rows. Calyx 6–8 (10) mm long ...........
..................................................... 6. A. lycopodioides (Girard.) Boiss.

+ Leaves acicular, 0.5–1 mm thick, divaricate. Peduncles 2.5–3.5 cm tall, with capitate 1-sided inflorescence bearing 4 spikelets. Calyx 11–12 cm long ..................................... 7. A. popovii Czern.

1. A. alatavicum Bge. in Mem. Ac. Sci. St.-Petersb. 7 ser., 18, 2 (1872) 40, p.p. excl. syn. Rupr.; Danguy in Bull. Mus. nat. hist. natur. 20 (1914) 72; Lincz. in Fl. SSSR, 18 (1952) 356; Fl. Kirgiz. 8 (1959) 157); Fl. Kazakhst. 7 (1964) 56; Peng in Fl. Sin. 60, 1 (1987) 17, incl. var. *laevigatum* Peng; Lincz. in Opred. rast. Sr. Azii [Key to Plants of Mid. Asia] 10 (1993) 26.— *A. kaschgaricum* Lincz. in Novit. syst. pl. vasc. 17 (1980) 209.

—Ic.: Fl. Sin. 60, 1, tab. 3, fig. 9–10.

Described from East. Kazakhstan (Jung. Ala Tau and vicinity of Issyk-kul' lake). Type in St.-Petersburg (LE). Map 4.

On arid rocky slopes of mountains, arid steppes and wormwood groves, deserts, 1200–3000 m.

IB. Kashgar: *Nor.* (nor.-west. fringe of Baisk basin, Adyr mountain range on left bank of Muzart river, desert belt, along gorge, Sept. 4; same site, south. extremity of old Muzart moraine, along gorge, Sept. 12–1958, Yun. et Yuan'; Uch-Turfan, Kukurtuk gorge, desert, May 31, 1908—Divn.).

IIA. Junggar: *Jung. Alat.* ("montgnes entre 1'Ebi-Nor et 1'Irtich, 1900 m, Aug. 20, 1895, Chaff." —Danguy, l.c.; oberstes Dunde-Kelde Tal und oberstes Chustai Tal, Aug. 5–6, 1908—Merzb.; "Wenquan, No. 4564, 1957, K.C. Kuan"—Peng, l.c.), *Tien Shan* (Sairam lake, July 20–24, 1877; Khanakhai, 1500–2150 m, June 16, 1878—A. Reg.; Uital river gorge, 2300 m, [June 1] 1889—Rob.; decliv. australis, beim Aul Kaitsche auf den Schutthügeln der Steppe, gegen Mitte, June 1903; decliv. australis, oberstes Dschanart Tal bis hinauf zum Pass, July 14-17, 1903—Merzb.).

General distribution: Jung.-Tarb., Nor. and Cent. Tien Shan; Mid. Asia (nor. Pam.-Alay).

2. **A. borodinii** Krassn. Spisok rast. Vost. Tyan'-Shanya [List of Plants of East. Tien Shan] (1886) 93, 128; Czern. in Tr. Bot. inst. AN SSSR, ser. 1, 3 (1937) 266; Lincz. in Fl. SSSR, 18 (1952) 317; Fl. Kirgiz. 8 (1959) 144; Fl. Kazakhst. 7 (1964) 51; Peng in Fl. Sin. 60, 1 (1987) 17; Lincz. in Opred. rast. Sr. Azii [Key to Plants of Mid. Asia] 10 (1993) 15.— *A. diapensioides* var. *borodinii* Krassn. in Zap. Russk. Geogr. Obshch. 19 (1888) 340.— *A. roborowskii* Czern. in Acta Inst. bot. Ac. Sci. URSS, ser. 1, 3 (1937) 267.

—Ic.: Fl. Sin. 60, 1, tab. 3, fig. 5–6.

Described from Sinkiang (East. Tien Shan). Type in St.-Petersburg (LE).

In melkozem-rocky and loessial mountain slopes in mountain-steppe and alpine belts, 2100–3000 m.

IIA. Junggar: *Tien Shan* (south. slope: Prope fl. Bedel non procul a castello chinensium, 1885 – Krassnov, typus!; Ui-tal river gorge, 2150–2450 m, along loessial descents of mountains, grows in clumps up to 3 ft in diam., June 1, 1889—Rob., typus *A. roborowskii* Czern.; Bei der Alpe oberhalb Sabantschö Gletscher, June 22–24; Oberstes Dschanart Tal bis hinauf zum Pass, July 14–17–1903, Merzb.).

General distribution: Cen. Tien Shan.

3. **A. diapensioides** Boiss. in DC. Prodr. 12 (1848) 624; ejusd. Fl. or. 4 (1879) 830; Alcock, Rep. Natur. Hist.-Results Pamir Bound. Commiss. (1898) 24; Czern. in Tr. Bot. inst. AN SSSR, ser. 1, 3 (1937) 258; Lincz. in Fl. SSSR, 18 (1952) 318; Fl. Kirgiz. 8 (1959) 150; Ikonnik. Opred. rast. Pamira [Key to Plants of Pamir] (1963) 194; Peng in Fl. Sin. 60, 1 (1981) 19; Lincz. in Opred. rast. Sr. Azii [Key to Plants of Mid. Asia] 10 (1993) 15.

—Ic.: Czern. l.c. fig. 1; Fl. SSSR, 18, Plate 17, fig. 3; Fl. Sin. 60, 1, tab. 3, fig. 3–4.

Described from Afghanistan (Kukhi-Baba mountain range). Type in London (K), isotype in St.-Petersburg (LE).

Along gently sloping melkozem slopes and floors of valleys, along banks of lakes, in high mountains, 2700–4800 m.

IIIC. Pamir (?).

General distribution: East. Pam.; Fore Asia (east. Afghanistan).

Note. We did not have specimens from Sinkiang.

4. A hedinii Ostenf. in Hedin, S. Tibet, 6, 3 (1922) 48; Czern. in Tr. Bot. inst. AN SSSR, ser. 1, 3 (1937) 260; Lincz. in Fl. SSSR, 18 (1952) 317; Fl. Kirgiz. 8 (1959) 149; Ikonnik. Opred. rast Pamira [Key to Plants of Pamir] (1963) 196; Peng in Fl. Sin 60, 1 (1987) 19; Lincz. in Opred. rast. Sr. Azii [Key to Plants of Mid. Asia] 10 (1993) 15.— A. diapensioides var. longifolia O. Fedtsch. in Acta Horti Petrop. 21 (1903) 407.— A. tianschanicum Czern. in Acta Inst. bot. Ac. Sci. URSS, ser. 1, 3 (1937) 262; Lincz. in Fl. SSSR, 18 (1952) 319; Fl. Kirgiz. 8 (1959) 150; Ikonnik. Opred. rast. Pamira [Key to Plants of Pamir] (1963) 196; Peng in Fl. Sin. 60, 1 (1987) 21; Lincz. in Opred. rast. Sr. Azii [Key to Plants of Mid. Asia] 10 (1993) 16.

—Ic.: Ostenf. l.c. pl. IV, fig. 2; Czern. l.c. fig. 2 and 3; Fl. Sin. 60, 1, tab. 3, fig. 1–2 (sub nom. A. tianschanicum).

Described from Sinkiang (Pamir). Type in Copenhagen (C). Plate I, fig. 5; map 2.

On melkozem-rocky slopes, old moraines, rock talus in alpine steppes and deserts, 2500–4700 m.

IB. Kashgar: Nor. (valley of Muzart river in Oi-Terek area, small passage on right in Tupu-daban creek valley, 2700 m, wormwood-cereal grass steppe, Sept. 7; same site, moraine on right flank of Tupu-daban creek valley, wormwood-saltwort steppified desert, Sept. 11; 10–12 km south-east of Akchit hydraulic station; nor.-west. slope of Sogdyn-tau mountain range, 2600 m, wormwood grove, Sept. 19–1958, Yun. and Yuan'), West. (Sulu-Sakal valley, 25 km east of Irkeshtam, 2900–3000 m, July 26, 1935 – Olsuf'ev; Kizil-daban pass, 3300 m, July 9, 1941 – Serp.; nor. slope of Kunlun, Akkëz-daban pass 115–120 km south of Kargalyk on Tibet highway, 3200 m, steppe belt, May 31; 32 km east of Ulugchat settlement on road to Kensu mine, saltwort desert along badland slope, June 18–1959, Yun. and Yuan').

IIA. Junggar: Tien Shan (Uch-Turfan, Karau gorge spruce, June 22, 1908 – Divn.; 3 km south-west of Torugart on highway to Kashgar, 3600 m, alpine steppe, July 20, 1959 – Yun. and Yuan').

IIIC. Pamir ("Eastern Pamir, among mosses on Kara-kir, on the eastern shore of Little Kara-kul, 3720 m, July 16, 1894 – Hedin", typus !; Tagdumbasch-Pamir; in angustis Pistan jugi Sarykol in detritu lapidoso, circa 4250 m, Feb. 15; in valle Ilyk-su prope Sassykteke, 4100 m, in siccis, July 16–1901, Alexeenko; Ulug-tuz gorge in Charlym river basin, along descents, June 27; Muztag-ata foothills, on talus, July 20–1909, Divn.; waterdivide of Kail'cha river and brook on east, 5500–6000 m, July 13, 1942 – Serp.).

General distribution: Cent. Tien Shan, East. Pam.; Mid. Asia (Alay valley).

Note. There are no morphological differences whatsoever between *A. tianschanicum* Czern. and *A. hedinii* Ostenf. The "characteristics" pointed out by Czerniakovskaya for her species are individual differences among different plants within the range of natural variation and are not at all related to the colour of calyx. There are also no geographic differences between red- and white-flowered forms: plants with dark raspberry flowers are distributed in Eastern Pamir as well while white-flowered plants are found in Central Tien Shan. There is a whole range of transitional forms of varying intensities of pink-flowered plants between red- and white-flowered plants. A study of fairly abundant herbarium material showed that acute leaves and bracts, although characteristic of red-flowered. *A. tianschanicum*, are also common among Pamir white-flowered plants and, on the contrary, obtuse leaves and bracts among Tien Shan red-flowered variety. Similar is the status even in relation to fine villosity on leaf margin. One-, 2- and even 3-flowered spikelets are equally found among red- and white-flowered plants, and among Pamir and Tien Shan plants. A similar phenomenon of colour variation of calyx is also noticed among *Limonium bicolor* (Bge.) Ktze. in which it ranges from bright raspberry to pure white, even on the same plant.

13
5. A. kokandense Bge. ex Rgl. in Acta Horti Petrop. 3 (1875) 99; Lincz. in Fl. SSSR, 18 (1952) 359; Fl. Kirgiz. 8 (1959) 158; Peng in Fl. Sin. 60, 1 (1987) 18; Lincz. in Opred. rast. Sr. Azii [Key to Plants of Mid. Asia] 10 (1993) 24.

Described from Mid. Asia (Alay mountain range). Type in Moscow (MW). Isotype in St.-Petersburg (LE).

On rocky and rubbly desert and arid steppe slopes and mountain trails, 2000–2700 m.

IB. Kashgar: *West.* (Turkestania orientalis, July 16 and Aug. 11, 1889 – Gromb.; 17 versts – 1 verst = 1.067 km – from Irkeshtam, around mountains, Aug. 11, 1913 – Knorring; Sulu-Sakal valley, 25 km east of Irkeshtam, juniper zone, July 25, 1935 – Olsuf'ev; King-tau mountain range, nor. slope 1 km north of Kosh-Kulak settlement, steppe belt, June 10; same site, nor. foothills 1–2 km south-east of Kosh-Kulak settlement, steppe belt, June 10; 67–68 km west of Kashgar on road to Kensu mine and Ulugchat, along gently sloping ridge, 2300 m, steppified desert, June 17; 30 km east of Ulugchat settlement on road to Kensu mine, intermontane valley, steppified wormwood desert, June 18; 60 km from Ulugchat settlement on road to Kensu mine, 2650 m, on badland slopes, saltwort desert, June 18–1959, Yun. and Yuan').

General distribution: Mid. Asia (Alay valley).

6. A. lycopodioides (Girard.) Boiss. in DC. Prodr. 12 (1848) 632; Bge. in Mem. Ac. Sci. St.-Petersb. 7 ser. 18 (1872) 20, descr. emend.; Hook. f. Fl. Brit. Ind. 3 (1884) 479; Pamp. Fl. Corac. (1930) 169, p.p. ?; Persson in Bot. notiser (1938) 296; Lincz. in Bot. mat. (Leningrad) 14 (1951) 278; id. in Fl. SSSR, 18 (1952) 314; Ikonnik. Opred. rast. Pamira [Key to Plants of Pamir] (1963) 196; Li Hen in Fl. Xizang, 3 (1986) 851; Lincz in Opred. rast. Sr. Azii [Key to Plants of Mid. Asia] 10 (1993) 16. – *Statice lycopodioides* Girard. in Ann. sci. natur. 3, ser. 2 (1844) 330, excl. syn. Willd. – *Acantholimon tibeticum* Hook. f. et Thoms. ex Bge. l.c. 20, nomen.

Described from Himalayas (Upper Indus river basin). Type in Vienna (W). Isotype in St. Petersburg (LE).

On rocks and rocky slopes in alpine belt.

IIIB. Tibet: *Chang Tang* ("nor.-west., Kunlun" – Fl. Xizang, l.c.).

IIIC. Pamir ("Jersil, 3400 m, July 14, 1930" – Persson, l.c.; Tagdumbasch Pamir; ad juitionem fl. Kara-czukur et Ilyk-su secus fl. Kara-czukur, loco Beikuy-auzy, about 3950 m, in rupestribus, July 18, 1901 – Alexeenko).

General distribution: East. Pam.; Himalayas (east.).

7. A. popovii Czern. in Acta Inst. bot. Ac. Sci. URSS, ser. 1, 3 (1937) 264; Peng in Fl. Sin. 61, 1 (1987) 18.

Described from Sinkiang (Kashgar). Type in St.-Petersburg (LE).

IB. Kashgar: *West.* (between Shur-bulak village and Kandzhugan [Kizil-su basin], July 4, 1929 – Pop., typus !).

General distribution: endemic.

## 4. Ikonnikovia Lincz.
### in Fl. URSS, 18 (1952) 745

1. I. kaufmanniana (Rgl.) Lincz. in Fl. URSS, 18 (1952) 381, Fl. Kazakhst. 7 (1964) 60; Fl. Sin. 60, 1 (1987) 22; Lincz. in Opred. rast. Sr. Azii [Key to Plants of Mid. Asia] 10 (1993) 28. – *Statice kaufmanniana* Rgl. in Acta Horti Petrop. 6, 2 (1880) 300 and in Gartenfl. 29, (1880) 1.

– Ic.: Gartenfl. 29, tab. 996; Fl. SSSR, 18, Plate 19, fig. 3; Fl. Kazakhst. 7, Plate 7, fig. 6; Fl. Sin. 60, 1, tab. 4, fig. 4.

Described from Sinkiang (Ili river basin). Type in St.-Petersburg (LE). Plate IV, fig. 5.

On rocky slopes, rocks and talus in low mountains, up to 2500 m.

IIA. Junggar: *Tien Shan* (Berg Bogdo, 2150-2450 m, July 24; ad fl. Kasch, Aug. 20–1878, A. Reg.), *Dzhark.* (Bach Chanachai, 3000–4000 ft [900–1200 m], June 15, 1878 – A. Reg., typus !; Back Chanachai südw. v. Kuldscha, 4000–5000 ft [1200–1500 m], June 15, 1878 – A. Reg., paratypus !; fl. Chanachai, 4000 ft [1200 m], June 27, 1878 – A. Reg., paratypus !).

General distribution: Nor. Tien Shan (nor. slopes and foothills of Trans-Ili Ala Tau and Ketmen' mountain range).

## 5. Goniolimon Boiss.
### in DC. Prodr. 12 (1848) 632

1. Spikes compact, quite densely capitate .................................... 2.

+ Spikes loose, with spaced spikelets, on tall peduncles branched in upper half; leaves lanceolate to narrow-elliptical ....................
.......................................... 1. G. callicomum (C.A. Mey.) Boiss.

2. Peduncles invariably branched more than twice, angular or ribbed above (but not crisp-winged), almost invariably puberulent; spikes with rather small heads and rather narrow bracts ..................................................................... 3.

+    Peduncles branched not more than twice, orbicular or crisp-winged above and almost invariably long-, compact–haired; spikes large-, dense-capitate with broad-ovoid bracts .............. 4.

3.    Plants up to 30, rarely up to 50 cm tall; leaves 5–10, rarely up to 15 cm long ..................................... 5. G. speciosum (L.) Boiss.

+    Plants up to 80 cm tall; leaves 10–15, quite often up to 20 cm long ......................... 2. G. dschungaricum (Rgl.) O. et B. Fedtsch.

4.    Peduncles mostly crisp-winged above; leaves up to 10 cm long with petioles 1/3–1/2 of blade length, almost invariably ciliolate along margin. Bracts with very broad membranous border ......................................................... 4. G. orthocladum Rupr.

+    Peduncles orbicular above; leaves large, up to 20 cm long, long-petiolate, as long as blade or less than 1/2 of it, invariably glabrous and smooth. Bracts with fairly broad membranous border ........................................... 3. G. eximium (Schrenk) Boiss.

1. G. callicomum (C.A. Mey.) Boiss. in DC. Prodr. 12 (1848) 633; Kryl. Fl. Zap. Sib. 9 (1937) 2163; Lincz. in Fl. SSSR, 18 (1952) 394; Fl. Kazakhst. 7 (1964) 68; Grub. Opred. rast. Mong. [Key to Plants of Mongolia] (1982) 199; Peng in Fl. Sin. 60, 1 (1987) 25; Lincz. in Opred. rast. Sr. Azii [Key to Plants of Mid. Asia] 10 (1993) 31. — *Statice callicoma* C.A. Mey. in Bong. et Mey. Verzeichn. Saisang-Nor Pfl. (1841) 56 and in Bull. Ac. Sci. St.-Petersb. 8 (1841) 340; Ldb. Fl. Ross. 3, 1 (1849) 465.

— Ic.: Fl. Sin. 60, 1, tab. 4, fig. 1.

Described from East. Kazakhstan (Zaisan lake region). Type in St.-Petersburg (LE).

On sandy, rocky-sandy and rubbly trails and slopes of mountains and conical hillocks.

IA. Mongolia: *Mong. Alt.* (Altyntschetsche, Iter ad Kobdo, June 22, 1870 — Kalning).

IIA. Junggar ("west. and nor.-west. Junggar" — Peng, l.c.).

General distribution: Aralo-Casp. (east.), Fore Balkh., Jung.-Tarb. (foothills), Nor. Tien Shan (Kungei-Ala Tau); Mid. Asia (West. Tien Shan and Karatau foothills), West. Sib. (upper Irtysh and Zaisan lake, Altay — Chui steppe).

Note. We did not study specimens of this species from Sinkiang; Komarov Botanical Institute does not have them in its herbarium.

2. G. dschungaricum (Rgl.) O. et B. Fedtsch. Consp. fl. Turkest. 5 (1913) 179; Lincz. in Fl. SSSR, 18 (1952) 390; Fl. Kazakhst. 7 (1964) 65; Peng in Fl. Sin. 60, 1 (1987) 25; Lincz. in Opred. rast. Sr. Azii [Key to Plants of Mid. Asia] 10 (1993) 30. — *G. tarbagataicum* Gamajun. in Vestn. AN Kazakhsk. SSR, 1 (1951) 80. — *Statice dschungarica* Rgl. in Acta Horti Petrop. 6, 2 (1880) 386.

— Ic.: Fl. Kazakhst. 7, Plate 8, fig. 3.

Described from East. Kazakhstan (Jung. Ala Tau). Type in St.-Petersburg (LE).

On rocky and meadowy steppe slopes in midbelt of mountains.

IIA. Junggar: *Tarb.* ?, *Jung. Alat.* ?, *Tien Shan* (Sairam lake, June 24; Talki river gorge, 1850–2150 m, July 25–1877, A. Reg.; on Sairam lake, July 23, 1878 – Fet.; upper Khorgos, 2450 m, Aug. 1878 – A. Reg.).

General distribution: Jung.-Tarb.

3. G. eximium (Schrenk) Boiss. in DC. Prodr. 12 (1848) 634; Lincz. in Fl. SSSR, 18 (1952) 391; Grub. Konsp. fl. MNR [Conspectus of Flora of Mongolian People's Republic] (1955) 221; Fl. Kirgiz. 8 (1959) 163; Fl. Kazakhst. 7 (1964) 66; Grub. Opred. rast. Mong. [Key to Plants of Mongolia] (1982) 198; Peng in Fl. Sin. 60, 1 (1987) 26, p.p.; Lincz. in Opred. rast. Sr. Azii [Key to Plants of Mid. Asia] 10 (1993) 30. – *Statice eximia* Schrenk in Fisch. et Mey. Enum. pl. nov. 1 (1841) 13; Ldb. Fl. Ross. 3, 1 (1849) 462.

– Ic.: Bot. Reg. 33, tab. 2; Fl. Kazakhst. 7, Plate 8, fig. 4; Grub. Opred. rast. Mong. [Key to Plants of Mongolia] Plants 107, fig. 480; Fl. Sin. 60, 1, tab. 4, fig. 2–3.

Described from East. Kazakhstan (Jung. Ala Tau). Type in St.-Petersburg (LE).

On rubbly and rocky steppe slopes in lower and middle belts of mountains.

IA. Mongolia: *Mong. Alt.* (zwischen d. Fluss. Khobdo und Hatu, June 20, 1870 – Kalning).

IIA. Junggar: *Tarb.* ?, *Jung. Alat.* ? *Tien Shan* ("Tien Shan and nor.-west. Junggar" – Peng, l.c.) ?

General distribution: Fore Balkh., Jung. Tarb., Nor. Tien Shan.

Note. Komarov Botanical Institute of the Russian Academy of Sciences does not have specimens of this species from Sinkiang in its herbarium.

4. G. orthocladum Rupr. in Mem. Ac. Sci. St.-Petersb. 7 ser. 14, 4 (1869) 69; Lincz. in Fl. SSSR, 18 (1952) 392; Fl. Kirgiz. 8 (1959) 161; Fl. Kazakhst. 7 (1964) 67; Lincz. in Opred. rast. Sr. Azii {Key to Plants of Mid. Asia] 10 (1993) 31. – *G. eximium* auct. non Schrenk; Peng in Fl. Sin. 60, 1 (1987) 26, pro syn. – *G. sewerzovii* Herd. in Bull. Soc. Natur. Moscou, 45, 1 (1872) 380, pp. quoad syn. Rupr.; Persson in Bot. notiser (1938) 296.

– Ic.: Fl. Kazakhst. 7, Plate 8, fig. 5.

Described from Cent. Tien Shan (Dzhaman-Daban). Type in St.-Petersburg (LE). Plate IV, fig. 4.

On rubbly and melkozem steppe slopes in middle and upper belts of mountains.

IIA. Junggar: *Jung. Alat.* ?, *Tien Shan* (on Talki river, 600–1800 m, 1874—Larionov; in Sumbe valley, 1850–2150 m, July 29, 1877—Fet.; Dzhagastai, 1500–2150 m, Aug. 9, 1877—A. Reg.; Shary-su river, 2150–2450 m, June 26, 1878—A. Reg.; on Syarmin river, Aug. 26, 1878—Larionov; upper Taldy river, 2450–2750 m, May 17; Irenkhabirga mountain range, Tsagan-usu river [Dzliin], 1850–2450 m, June 10; Naryn gol near Tsagan-usu, 1850–2450 m, June 10; Borgaty, 1850 m, July 4; Mëngëtë on nor. slope of Irenkhabirga, 2750 m, July 9; Maralty, Kunges, 2450 m, Aug.—1879, A. Reg.; Khaidyk-gol river, 2150–2450 m, Aug. 9, 1893—Rob.; Urumchi region, upper Tasenku river, Biangou area, Sept. 24, 1929—Pop.; "Kirgis-at davan, 2600 m, July 30, 1932"—Persson, l.c.).

General distribution: Jung.-Tarb. (Jung. Alat.), Nor. and cent. Tien Shan.

5. G. speciosum (L.) Boiss. in DC. Prodr. 12 (1848) 634; Gr.-Grzh. Zap. Mong. 3 (1930) 820; Kryl. Fl. Zap. Sib. 9 (1937) 2162; Lincz. in Fl. SSSR, 18 (1952) 388; Fl. Kazakhst. 7 (1964) 65; Fl. Tsentr. Sib. 2 (1979) 707; Fl. Intramong. 5 (1980) 45; Grub. Opred. rast. Mong. [Key to Plants of Mongolia] (1982) 199; Opred. rast. Tuv. ASSR [Key to Plants of Tuva Autonomous Soviet Socialist Republic] (1984) 79; Peng in Fl. Sin. 60, 1 (1987) 23; Lincz. in Opred. rast. Sr. Azii [Key to Plants of Mid. Asia] 10 (1993) 30.—*Statice speciosa* L. Sp. pl. 1 (1753) 275; Gr.-Grzh. Zap Kitai, 3 (1907) 497; Sap. Mong. Altai (1911) 383; Danguy in Bull. Mus. nat. hist. natur. 20 (1914) 73.—*Goniolimon strictum* (Rgl.) Lincz. in Fl. URSS, 18 (1952) 395, in not.; Fl. Kazakhst. 7 (1964) 69; Lincz. in Opred. rast. Sr. Azii [Key to Plants of Mid. Asia] 10 (1993) 31.—*Statice speciosa* var. *stricta* Rgl. in Acta Horti Petrop. 6, 2 (1880) 389, 387, in nota.—*Limonium speciosum* (L.) Ktze. Rev. Gen. pl. 2 (1891) 396; Kitag. Lin. fl. Mansh. (1939) 354.

—Ic.: Bot. Mag. 18, tab. 656; Fl. Kazakhst. 7, Plate 8, fig. 1; Fl. Intramong. 5, tab. 18, fig. 1–4.

Described from plants grown in Uppsala from seeds received from "Tataria" (Trans-Volga ?). Type in London (Linn.).

On rubbly and rocky steppe and desert slopes and trails of mountains, on pebble terraces of rivers and lakes, on rocks and crests of conical hillocks.

IA. Mongolia: *Khobd., Mong. Alt., Cent. Khalkha, East, Mong., Depr. Lakes, Valley Lakes, Gobi-Alt., East. Gobi, West. Gobi* (valley of Tukhumyin-Khundei, Aug. 9, 1947—Yun.), *Alash. Gobi* (Kobden-usa, Aug. 14, 1886—Pot.).

IIA. Junggar: *Tarb.* (Kotbukha mountains, rocks, Aug. 10, 1876—Pot.), *Jung. Alat.* (Maili mountain range, Daganbel' mountains 4 km south of Dzhirmasu picket, 1300 m, July 18, 1953—Mois.), *Tien Shan* (in lower Kunges river valley, 900–1050 m, Sept. 2, 1876; same site, about 750 m, July 30, 1877—Przew.*; Aktyube north of Kul'dzha, May 13*; Pilyuchi near Kul'dzha, May 17*; Sairam lake, Talki brook, July 19; Talkibash mountains on Sairam lake, July 20; Talki gorge, July; Sairam lake, south-east. bank, July 23–1877, A. Reg.; Sairam-Nor valley, July 12; near Sairam lake, July 19; west of Sairam lake, July 20; Urtak-sary [Sairam], July 20–1878, Fet.; Kok-kamyr mountains, 1850–2150 m, July 27, 1878; Bainamun near Dzhin river, 1500–1850 m, June 5, 1879—A. Reg.; south of Dzhin-kho, between Dzhus-agach and Boro-Khoro mountains, June 15, 1889—Gr.-Grzh.; "Sairam-Nor, July 24, 1895, Chaff."—Danguy, l.c.; Kapsalyon Tal und Nebental

Kisyl-sai, auch Plateauhöhe von Karadschon, Aug. 2–3, 1907 — Merzb.\*; Dzin'kho, 10 km south of Sairam lake, knolls, Aug. 31, 1959; basin of Bol. Yuldus 3–5 km south-west of Bain-Bulak settlement, mountain on right bank of Khaidyk-gol, Aug. 10, 1958 — Yun. and Yuan'), *Jung. Gobi* (22–23 km west of Bidzhiin-gol on road to Bulgan somon, July 18, 1947 — Yun.; left bank of Chern. Irtysh 38 km east of Shipati crossing on road to Koktogai, hummocky area, July 8\*; 60 km south of Ertai settlement on road to Guchen, hummocky area, July 16–1959, Yun. and Yuan'; spurs of Argalante mountains 4 km from Ubchu-bulak, about 1700 m, sandy valley, July 2; Serteng-ula 20 km from Altay somon, about 1600 m, slopes, July 2–1973, Golubkova and Tsogt; Khoni-Usuni-khoolai area 100 km south-east of Bulgan somon, Aug. 27, 1973 — Isach. and Rachk.), *Zaisan* (Asy-Sary-bulak area, Aug. 5, 1876 — Pot.; left bank of Chern. Irtysh, 17–18 km south-west of Burchum settlement on road to Zimunai, July 10, 1959 — Yun. and Yuan').

General distribution: Aralo-Casp. (nor.), Fore Balkh. (nor.), Jung.-Tarb., Nor. Tien Shan; Europe (south. Trans-Volga and Fore Urals), West. Sib. (south.), East. Sib. (south.), Nor. Mong.

17     Note. Var. *strictum* (Rgl.) Peng [asterisked (\*)] l.c., characterised by very loose inflorescence, somewhat resembling the inflorescence of *Goniolimon callicomum* (C.A. Mey.) Boiss. is sporadically found in low mountains and foothills of Junggar Ala Tau, Nor. and East. Tien Shan and in Junggar Gobi. It differs very little from the parent species and is possibly of hybrid origin from the above species.

## 6. Armeria Willd.
### Enum. pl. Horti berolin. (1809) 339

1. A. sibirica Turcz. ex Boiss. in DC. Prodr. 12 (1848) 678; Ldb. Fl. Ross. 3, 1 (1849) 457; Turcz. in Bull. Soc. Natur. Moscou, 25, 3 (1852) 399; Shteinb. in Fl. SSSR, 18 (1952) 409; Grub. Konsp. fl. MNR [Conspectus of Flora of Mongolian People's Republic] (1955) 221; id. Opred. rast. Mong. [Key to Plants of Mongolia] (1982) 199. — *A. scabra* auct. non Pall.: Fl. Tsentr. Sib. 2 (1979) 708.

— Ic.: Fl. dan. 16, f. 2769.

Described from East. Siberia. Type in St.-Petersburg (LE).

In moist and marshy meadows and *Cobresia* groves, moist rubbly slopes and turf-covered rock formations, along banks of brooks in high mountains.

IA. Mongolia: *Mong. Alt.* (Taishiri-Ula, in ravine, July 13, 1877 — Pot.; same site, gently sloping nor.-east. Slopes near summit, Aug. 16 and on nor. slope near crest, Sept.. 21–1945, Leont'ev; slope of Shadzagain-Suburga pass, July 22, 1898 — Klem.; along islands of lower Kobdo lake in upper course of river, Aug. 2, 1899 — Lad.).

General distribution: East. Sib. (east. Sayan, east. Fore Baikal), Nor. Mong. (cent. Hangay).

Note. Binomial *Armeria scabra* Pall. ex Roem. et Schult. [Syst. veget. (1820) 776] is not applicable to this species since its protologue showed that leaves are crispate above, scape scabrous and outer phyllary oblong and cuspidate, while leaves and scape in *A. sibirica* are glabrous and smooth and phyllary oval and obtuse. ("*A. scabra* Pall.: scapo tereti scabriusculo, calycis foliolis exterioribus oblongis mucronatis, interioribus obtusis, foliis linearibus supra subpilosis, Reliqu. Willd. MS. In Asia boreali, Pall.".) This diagnosis corresponds more to *A. arctica* (Cham.) Wallr. but not to *A. sibirica*.

## 7. Limonium Mill.
### Gard. Dict. Abridg. 4 (1754)

1. Herbaceous perennials, sometimes with lignifying short caudex .................................................................................... 2.

+ Subshrubs with leafy lignifying ascending shoots and clusters of linear-lobed fleshy leaves at their tips. Flowers small, violet-coloured, in capitate racemes, interrupted-spicately sessile at tips of simple or poorly branched peduncles (section 2. Sarcophyllum (Boiss.) Lincz.). Leaves with a pair of auriculate membranous processes on petiole base ........................................ ............................................. 5. L. suffruticosum (L.) Ktze.

2. Calyx tubular with poorly manifest limb (section 3. Siphonocalyx Lincz.). Spikelets aggregated in compact one-sided spikes, arranged 1–3 at tips of inflorescence branches ....... ............................................. 6. L. callianthum (Peng) Grub.

+ Calyx infundibular or obconical with distinctly manifest limb .................................................................................... 3.

3. Calyx broad-infundibular, 4.5–12 mm long, with 2–7 mm broad limb, highly tapered at base (section 4. Platymenium Boiss.) ... ............................................. 4.

+ Calyx narrow-infundibular or obconical, 2.5–4 mm long, with 1–1.5 mm broad narrow limb, almost without a taper at base (section 1. Limonium) ................................................. 20.

4. Stems in lower nodes with 1–5 small green leaves; radical leaves very large, persist until end of vegetation. Flowers and inflorescence lilac-pink or whitish ................................................. 5.

+ Stems invariably without leaves, only with membranous glumes in nodes. Flowers and inflorescence yellow, purple, lilac, pink or white. Radical leaves withering early and usually caducous ................................................................................. 6.

5. Stems erect, 5–25 cm tall, usually single, rarely 2–5. Spikelets in globose-capitate inflorescences at end of branches; corolla lilac-pink or pink; calyx 6–7 mm long ...... 13. L. flexuosum (L.) Ktze.

+ Stems 5–20, procumbent, rising at tip, 3–10 cm long. Spikelets in small whorls; corolla pale yellow, calyx 3–4.5 mm long ...... ............................................. 10. L. congestum (Ledeb.) Ktze.

6. Plants forming caudex with silvery white scarious glumes and remnants of petioles at tips; stems usually several, erect or somewhat flexuose, their barren twigs short, erect and mostly simple ................................................................................. 7.

18

+    Plant not forming caudex; root tip with only dark brown remnants of dead leaves; stems single or many, but then angular-flexuose and highly branched, with abundant barren twigs ............................................................................ 14.

7.    Corolla pink-violet; limb of calyx lilac, pinkish (like inflorescence as a whole) or white, specially when dry ........... 8.

+    Corolla bright yellow and limb of calyx (like inflorescence as a whole) lemon-yellow ................................................................. 11.

8.    Flowers in loose inflorescences; calyx lilac. Stems thin and slender. Leaves linear-lobed ......................................................... 9.

+    Flowers in compact, subcapitate terminal inflorescences forming quite compact corymb; calyx pinkish or whitish. Stems strong with short internodes. Leaves linear, thickish ...................
.......................................................... 14. L. gobicum Ik.-Gal.

9.    Stems 2–3 (5), sometimes single, very thin; caudex small, up to 1–1.5 cm in diam. Calyx pinkish violet, not albescent until end of vegetation. Root bark highly dehiscent, root fibrous ............
....................................... 23. L. tenellum (Turcz.) Ktze.

+    Stems invariably many; caudex large, often multicipital, 2–7 cm in diam. Calyx light pinkish violet. Root entire, glabrous ...... 10.

10.    Calyx (7) 8–11 mm long, with long acute-deltoid lobes, fading rapidly to white; bracts invariably glabrous. Plant 10–25 cm tall, stems slender and quite thin ...............................................
.............................................. 17. L. kaschgaricum (Rupr.) Ik.-Gal.

+    Calyx 5–6 (7) mm long, with short rounded lobes, retaining colour until end of vegetation; bracts glabrous or (rarely) pubescent. Plant 3–10 (15) cm tall, stems thickish and uneven ....
........................................................ 16. L. hoeltzeri (Rgl.) Ik.-Gal.

11.    Leaves withering early and turning brown, narrow- or linear-lobed; glumes at base and nodes of stems and branches white-scarious, long, linear-lanceolate; barren twigs short, usually simple and sessile in clusters of 2–5 each or almost absent. Bracts whitish ........................................................................... 12.

+    Leaves long-persistent, green, broad-lobed, up to 3 cm long, 1 cm broad, obtuse, with fine notch at tip or with short implanted cusp; glumes in nodes of stems and branches, scarious, but with thick brown or greenish nerve, short, ovoid-lanceolate; barren twigs long and branched. Bracts mostly reddish ............
........................................................ 18. L. klementzii Ik.-Gal.

12.    Stems rather short (3–15 cm tall), branched only in upper third, with short internodes implanted with silvery white scarious

glumes in nodes, with short barren twigs, often in clusters. Flowers bright lemon-yellow, in capitate inflorescence ........ 13.

+ Stems thin, slender, glabrous, up to 35 cm tall, branched almost from base, with long internodes and mostly without barren twigs, with poorly visible glumes in nodes. Flowers sulphureous, in diffuse inflorescence ........................................
......................................................................... 21. L. roborowskii Ik.-Gal.

13. Stems and branches, at least at tip, verruculose under inflorescence; outer bracts of spikelets glabrous, rarely (in the eastern range of species) pubescent (var. *pubescens* Lincz.)
............................................. 9. L. chrysocomum (Kar. et Kir.) Ktze.

+ Stems and branches glabrous; outer and inner bracts of spikelets almost invariably compact and crinite, very rarely (in the western range of species) glabrous ....................................
......................................................... 22. L. semenovii (Herd.) Ktze.

14. Stems many, flexuose and mostly procumbent-ascending, 5–20 cm tall, highly branched from base; their barren twigs many, angular-flexuose and repeatedly branched, verruculose ...... 15.

+ Stems usually single or 2–3, very rarely more together, straight, erect, up to 60 cm tall, branched in upper half, with rather few erect, glabrous barren twigs ......................................................... 17.

15. Calyx (like inflorescence as a whole) bright golden yellow, corolla orange-coloured. Root bark dark brown ..................... 16.

+ Calyx (like inflorescence as a whole) white or pinkish, corolla yellow. Root bark beet-red ............ 12. L. erythrorhizum Ik.-Gal.

16. Stems and branches pubescent with stellate hairs; barren twigs very many in lower half of plant, highly branched, crispately into slender twigs. Stems erect, 25–40 cm tall. Plant of rocks and rocky semidesert slopes of mountains ........................................
....................................................................... 20. L. potaninii Ik.-Gal.

+ Stems and branches glabrous; barren branches not fine-crispately branched. Stems more often procumbent-ascending, 5–20 cm tall. Plant of solonchaks and saline soils .......................
.................................................................... 7. L. aureum (L.) Hill.

17. Stems thin, slender, rounded, with rather few, usually simple barren twigs. Leaves narrow-lobed, 0.5–1 cm broad. Inflorescence small, paniculate, loose; flowers fade rapidly and turn white ................................................................................. 18.

+ Stems or stem rough, strong, faceted, highly branched in inflorescence, barren twigs branched. Leaves oblong-obovoid, 1–2 cm broad. Inflorescence compact, corymbose or loose, broad-paniculate ......................................................................... 19.

18.    Calyx light pink-violet, (6) 7–9 mm long, its limb with acute-deltoid lobes, with awnlike cusp; outer bract of spikelet 1.5–3 mm long, 1/4–1/2 of first inner bract, latter invariably glabrous; corolla yellow ................................ 19. L. leptolobum (Rgl.) Ktze.

\+    Calyx pale yellow or whitish, 5–7 mm long, its limb with short orbicular lobes; outer bract of spikelet 3–4 mm long, 1/2 of first inner bract, latter more or less compact and crinite on back; corolla yellow .................... 11. L. dichroanthum (Rupr.) Ik.-Gal.

19.    Calyx (and entire inflorescence) bright purple to light pink and white; corolla yellow or colourless ....... 8. L. bicolor (Bge.) Ktze.

\+    Calyx (and inflorescence) golden yellow, corolla orange-yellow ............................................. 15. L. grubovii Lincz.

20.    All leaves exclusively radical; seen very rarely on lower part of stems but then leaf form same as that of radical leaves, petiolate ...................................................................... 21.

\+    Radical leaves obovoid-lobed, petiolate; orbicular-reniform or orbicular, sessile, amplexicaul on stem and lower part of branches. Flowers on short spikes, aggregated in compact panicle .............................. 4. L. otolepis (Schrenk) Ktze.

21.    Leaves 10–25 cm long, 5–10 cm broad, not perishing after anthesis. Peduncles usually rather few ...................................... 22.

\+    Leaves 1–3 (6) cm long, 0.5–2 cm broad, usually perishing by anthesis, peduncles mostly several (up to 20), 20–50 cm tall, with brown glumes implanted compactly at base, repeatedly branched from base and with many dichotomously branching barren twigs; inflorescence corymbose-paniculate, compact. Entire plant compact and short-hispid ......................................
................................................ 1. L. coralloides (Tausch) Lincz.

22.    Leaves subovoid to elliptical and obovoid, with single longitudinal nerve. Inflorescence compressed, subcorymbose or pyramidal; flowers on fairly compact short spikelets ...............
................................................ 2. L. gmelinii (Willd.) Ktze.

\+    Leaves broad-obovoid or broad-lobed to obdeltoid, with 3–5 longitudinal nerves. Inflorescence diffuse, loose-paniculate; flowers on long, rather loose spikelets ...............................................
................................................ 3. L. myrianthum (Schrenk) Ktze.

## Section 1. Limonium

1. L. coralloides (Tausch) Lincz. in Fl. URSS, 18 (1952) 451; Grub. Konsp. fl. MNR [Conspectus of Flora of Mongolian People's Republic] (1955) 221; Fl. Kazakhst. 7 (1964) 82; Grub. Opred. rast. Mong. [Key to

20

Plants of Mongolia] (1982) 119; Opred. rast. Tuv. ASSR [Key to Plants of Tuva Autonomous Soviet Socialist Republic] (1984) 79; Peng in Fl. Sin. 60, 1 (1987) 43; Lincz. in Opred. rast. Sr. Azii [Key to Plants of Mid. Asia] to (1993) 38. — *Statice coralloides* Tausch. Sylloge Ratisb. 2 (1828) 255. — *S. aphylla* Poir. Encycl. method. 7 (1806) 408, non Forsk. (1775); Kryl. Fl. Zap. Sib. 9 (1937) 2158. — *S. decipiens* Ldb. Fl. alt. 1 (1829) 433, p.p.; Sap. Mong. Altai (1911) 383; Danguy in Bull. Mus. nat. hist. natur. 20 (1914) 73.

— Ic.: Fl. Kazakhst. 7, Plate 9, fig. 4; Opred. rast. Tuv. ASSR [Key to Plants of Tuva Autonomous Soviet Socialist Republic] fig. 69; Fl. Sin. 60, 1, tab. 7, fig. 3–6.

21    Described from Siberia. Type in Praha (PR) ?

In solonchaks, solonetzes, solonetzic meadows, in chee grass thickets, along banks of saline lakes.

IA. Mongolia: *Depr. Lakes* (south. bank of Khara-Usu lake, Aug. 18, Ubsa lake, Sept. 22–1879, Pot.; on nor. bank of Ubsa lake, July 3, 1892 — Kryl.; Ubsa-nur, around Baga-nur lake, Sept. 6, 1895 — Klem.; Ubsa-nur, in Kharkhira river valley, Aug. 19, 1930 — Bar.; Ulangoma valley, Aug. 19, 1931 — Bar. and Shukhardin; nor. of Ulangoma town, Aug. 19, 1931 — Bar.).

IIA. Junggar: *Tarb.* (Mukurtai area west of Ulyungur lake, June 21, 1908 — Sap.), *Jung. Gobi* (Thal des Fl. Urungu, June 10–22, 1876 — Pewz.; in Shara-bulak brook valley, Aug. 4, 1898 — Klem.; "entre 1'Ouchte et l'Irtich, steppes, 810 m, Aug. 27, 1895, Chaff." — Danguy, l.c.; Dzhirgalantu river valley, Sept. 16, 1930 — Bar.; Bulgan river valley 5–7 km below Ulyasteam creek valley, July 28, 1947 — Yun.; ? No. 10367, June 4, 1959 — Lee et Chu; Urungu river valley 40–45 km below Ertai settlement on road from Din'syan, July 13, 1959 — Yun. and Yuan'), *Zaisan* ([right bank of Chern. Irtysh below Burchum river] Sary-dzhasyk — Kiikpai well, June 15, 1914 — Schischk.).

General distribution: Fore Balkh.; West. Sib. (Irt. south, Altay south-west), East. Sib. (Tuva basin south).

2. L. gmelinii (Willd.) Ktze. Rev. gen. 2 (1891) 395; Lincz. in Fl. SSSR, 18 (1952) 436; Grub. Konsp. fl. MNR [Conspectus of Flora of Mongolian People's Republic (1955) 222; id. in Bot. mat. (Leningrad) 19 (1959) 548; Fl. Kirgiz. 8 (1959) 169; Fl. Kazakhst. 7 (1964) 79; Fl. Tsentr. Sib. 2 (1979) 709; Grub. Opred. rast. Mong. [Key to Plants of Mongolia] (1982) 199; Opred. rast. Tuv. ASSR [Key to Plants of Tuva Autonomous Soviet Socialist Republic] (1984) 79; Peng in Fl. Sin. 60, 1 (1987); Lincz. in Opred. rast. Sr. Azii [Key to Plants of Mid. Asia] 10 (1993) 38. — *Statice gmelinii* Willd. Sp. pl. 1 (1797) 1524; Gr.-Grzh. Zap. Kitai, 3 (1907) 497; Sap. Mong. Altai (1911) 383; Danguy in Bull. Mus. nat. hist. natur. 20 (1914) 73; Kryl. Fl. Zap. Sib. 9 (1937) 2154; Chen et Chou, Rast. pokrov r. Sulekhe [Vegetation Cover of Sulekhe river] (1956) 81.

— Ic.: Fl. Kazakhst. 7, Plate 9, fig. 1; Fl. Sin. 60, 1, tab. 7, fig. 7–9.

Described form Siberia. Type in Berlin (B).

In arid solonchaks, solonchaklike and solonetzic meadows, in chee grass thickets, saline sinkholes in deserts and steppes, along banks of saline lakes and desert rivers.

IA. Mongolia: *Khobd.* (east. fringe of Achit-nur basin, Ulyasutuin-gol valley, Aug. 7, 1947 — Tarasov), *Depr. Lakes* (Ubsu-nur basin: Kundelen river, Sept. 20, 1879 — Pot.; same site, around Baga-nur lake, Sept. 6, 1895 — Klem.; same site, north of Ulangom, Aug. 19, 1931 — Bar.; same site, Borig-Del' sand, Khurmusin-gol 22 km west of Dzun-Gobi somon centre, July 22, 1971 — Grub., Dariima, Ulzij.).

IB. Kashgar: *West.* (Artush, 10 km south-east of Khalatsi, 1800 m, No. 9798, June 22, 1959 — Lee et Chu).

IIA. Junggar: *Cis-Alt.* (Vorberge den südlichen Altai, 1876 — Pewz.; steppe in Kemerchuk river valley, Aug. 15, 1906 — Sap.; 60 km south of Koktogai on road to Ertai on Urungu river, July 14, 1959 — Yun. et Yuan'; Kobulchi Altay, on brook alongside road, No. 10784, July 18, 1959 — Lee et Chu), *Tarb.* (Chuguchak basin 10 km south-east of Durbul'dzhin settlement on road to Toli, Aug. 8, 1957 — Yun., S.-y. Lee, Yuan'), *Jung. Alat.* (Dzhair mountain range, chee grass steppe on Shikho-Chipeitsza road, Aug. 18, 1952 — Mois.; 45 km north of Toli, No. 1415, Aug. 9, 1957 — Shen), *Tien Shan* (Talki gorge, July 25, 1877; on Kash river, Sept. 6, 1878; Kash river between Ulastai and Nilki, 900– 1200 m, June 30, 1979 — A. Reg.; 200 m north of Gunlyu, 900 m, No. 1074, July 20, 1957 — Shen; left bank of Kunges 5–8 km west-nor.-west of state farm on road at crossing, Aug. 29, 1957 — Yun., S.-y. Lee, Yuan'), *Jung. Gobi* (north of Guchen, Gashun', No. 23, 1889 — Gr.-Grzh.; between Khon'chyu and Dondkho, Aug. 12; between Dzhimuchi and Dachuan, Aug. 21 — 1898, Klem., Ebi-nor lake, about 1950 m, Aug. 27, 1953 — Mois.; along bank of Manas river, 27 km north-west of Paotai settlement, No. 134, June 17, 1957 — Shen; on road to 16th state farm and Urumchi from Khutubi, 600 m, gobi, No. 5071, Sept. 21, 1957 and ? No. 4820, Sept. 2, 1954 — Kuan; valley of Chumpazy river, right bank 2–3 km from Arasan on road to Temirtam, Aug. 6, 1957 — Yun., S.-y. Lee, Yuan'), *Zaisan* (nor. bank of Chern. Irtysh, opposite Cherektas mountains, June 4; same site Maikain area, June 7–1914, Schischk.), *Dzhark.* (on Kapshachai river [near Kul'dzha], 360–1800 m, 1874 — Larionov; Khoir sumun south of Kul'dzha, May 27; Suidun town, July 8–1877, A. (Reg.).

General distribution: Aralo-Casp., Fore Balkh., Cent. Tien Shan; Europe (Cent., East. and Europ. Russia, south.), Mid. Asia (plains), West. Sib. (south.), East. Sib. (Ang.-Sayan).

3. **L. myrianthum** (Schrenk) Ktze. Rev. gen. 2 (1891) 395; Lincz. in Fl. SSSR, 18 (1952) 453; Fl. Kirgiz. 8 (1959) 170; Fl. Kazakhst. 7 (1964) 83; Sanczir, Rachk. et al. in Izv. AN MNR, ser. biol. 3 (1985) 52; Fl. Sin. 60, 1 (1987) 43; Lincz. in Opred. rast. Sr. Azii [Key to Plants of Mid. Asia] 10 (1993) 40. — *Statice myriantha* Schrenk in Fisch. et Mey. Enum. pl. nov. 1 (1841) 14; Kryl. Fl. Zap. Sib. 9 (1937) 2156; Persson in Bot. notiser (1938) 296.

— Ic.: Fl. Kazakhst. 7, Plate 9, fig. 5; Fl. Sin. 60, 1, tab. 7, fig. 1–2.

Described from East. Kazakhstan (Balkhash Alakul'sk region). Type in St.-Petersburg (LE).

In solonetzic meadows, chee grass thickets, tugais, along banks of desert rivers, springs and lakes.

IIA. Junggar: *Tien Shan* (Toguztarau at confluence of Tekes and Kunges rivers, Nov. 1876 — A. Reg.; along upper Ili, June 28; steppe near Kunges and upper Ili, June 18–July

3–1877, Przew.; "Taldi, about 2000 m, Aug. 25, 1932"—Persson, l.c.; intermontane Davanchin valley, Ulumbai oasis 20–25 km south of Urumchi, June 2; left bank of Kunges river 5–8 km west—nor.-west of state farm on road at crossing, Aug. 20–1957, Yun., S.-y. Lee, Yaun'), *Jung. Gobi* (lower Borotala, Uch'-tyube, Aug. 20; lower Borotala, 900 m, Aug. 22–1878, A. Reg.; between Khon'chyu and Dondkho, Aug. 12; between Dzhimuchi and Dachuan, Aug. 21–1898, Klem.; 15 km east of Kuitun river along highway, 20 km from Karamai settlement, No. 228, June 27; 6 km west of Ula-usu settlement, No. 317, July 3–1957, Shen; 18–20 km east of St. Kuitun settlement on road to Manas from Shikho, extensive solonchak lowland, July 7, 1957—Yuan., S.-y. Lee, Yuan'; 50 km east—nor.-east of Urumchi on road to Fukan, desert, April 26, 1959—Yun. et Yuan'; "Altay somon, Khonin-Usny shand, May 30, 1967—Shagdarsuren and Dovchin"—Sanczir, l.c.; 90 km south-east of Altay somon centre, Khonin-usu spring, 1420 m, Aug. 14, 1981—Rachk.), *Dzhark.* (bank of Ili river south-west of Kul'dzha, May 30; on Ili river, June 1–1877, A. Reg.; prope Ili non procul a Chorgos, July 1, 1886—Krassnov).

General distribution: Aralo-Casp. (south-east.), Fore Balkh., Dzhark.-Tarb., Cent. Tien Shan; Mid. Asia (nor. Syr-Dar.).

**4. L. otolepis (Schrenk) Ktze.** Rev. gen. 2 (1891) 396; Lincz. in Fl. SSSR, 18 (1952) 455; Grub. in Bot. mat. (Leningrad) 19 (1959) 548; Fl. Kirgiz. 8 (1959) 173; Fl. Kazakhst. 7 (1964) 85; Fl. Sin. 60, 1 (1987) 42; Lincz. in Opred. rast. Sr. Azii [Key to Plants of Mid. Asia] 10 (1993) 40.—*Statice otolepis* Schrenk in Bull. Ac. Sci. St.-Petersb. 1 (1843) 362; Gr.-Grzh. Zap. Kitai, 3 (1907) 497; Danguy in Bull. Mus. nat. hist. natur. 14 (1911) 9; Kryl. Fl. Zap. Sib. 9 (1937) 2159.

—Ic.: Fl. Kazakhst. 7, Plate 9, fig. 7; Fl. Sin. 60, 1, tab. 8, fig. 1–3.

Described from Cent. Kazakhstan (Betpak-dala). Type in St.-Petersburg (LE).

On solonchaks, solonchaklike meadows, in chee grass thickets and tugais, in sandy saline depressions between mounds, along margins of puffed solonchaks, along banks of irrigation ditches.

IA. Mongolia: *West. Gobi* (Beishan', Sy-dun well, Oct. 31, 1890—Gr.-Grzh.), *Alash Gobi* (35 km nor.-east of Tszin'ta town, solonchak on Beitakhe brook side, July 18, 1958—Petr.; 30 km nor.-east of Tszin'ta town, Edzin-gol river valley, Sept. 23, 1958—Lavr. et L.-i. Chen), *Khesi* (Suleikhe river valley between Yuimyn and An'si, Aug. 9 [21], 1875—Pias.; "Maraise de Lou-Tsao-keou entre Cha-Tsheau et Ngai-Si-Tsheau, July 11, 1908, Vaillant"—Danguy, l.c.; 24 km south-east of An'si town, in Tasykhe river valley, July 27, 1958—Petr.).

IB. Kashgar: *West.* (Yangishar kishlak—village in Central Asia—Aug. 3; same site, on road to Yangan kishlak, Aug. 3–1913, Knorring).

IIA. Junggar: *Jung. Alat.* (Dzhair mountain range, Tuz-agly ravine, oasis, June 10; Dzhair mountain range, on Shikho-Karamai road, Aug. 18–1951, Mois.), *Jung. Gobi* (left bank of Manas river 2–3 km nor.-west of Savan settlement on road to Paotai state farm, June 10, 1957—Yun., S.-y. Lee, Yuan'; 7 km nor.-east of Paotai settlement, on bank of Manas river, No. 68, June 11; 21 km nor.-west of Paotai settlement, on bank of Manas river, No. 128, June 17; in Paotai settlement region, No. 133, June 17; on bank of Kuitun river, No. 209, June 24–1957, Shen; ? No. 4820 and 4821, Sept. 2, 1957—Kuan; from Tszinchan town to 16th state farm, 700 m, gobi, No. 5044, Sept. 20, 1957—Kuan), *Dzhark.* (bank of Ili river south-west of Kul'dzha, May 30, 1877—A. Reg.; Chorgos, 1886—Krassnov).

IIIA. Qinghai: *Amdo* ("Yui-guan', Shidun' solonchak [Huang He valley], Sept. 8, 1890" — Gr.-Grzh., l.c.).

General distribution: Aralo-Casp., Fore Balkh.; Fore Asia (nor. Afghanistan), Mid. Asia (plains).

## Section 2. Sarcophyllum (Boiss.) Lincz.

5. L. suffruticosum (L.) Ktze. Rev. gen. 2 (1891) 396; Lincz. in Fl. SSSR, 18 (1952) 458; Grub. Konsp. fl. MNR [Conspectus of Flora of Mongolian People's Republic] (1955) 222; Fl. Kazakhst. 7 (1964) 86; Grub. Opred. rast. Mong. [Key to Plants of Mongolia] (1982) 199; Peng in Fl. Sin. 60, 1 (1987) 45; Lincz. in Opred. rast. Sr. Azii [Key to Plants of Mid. Asia] 10 (1993) 35. — *Statice suffruticosa* L. Sp. pl. (1753) 276; Ldb. Fl. Ross. 3 (1849) 468; Danguy in Bull. Mus. nat. hist. natur. 20 (1914) 73; Kryl. Fl. Zap. Sib. 9 (1937) 2160.

— Ic.: Fl. SSSR, 18, Plate 32, fig. 3; Fl. Kazakhst. 7, Plate 9, fig. 9; Fl. Sin. 60, 1, tab. 8, fig. 4–7.

Described from Siberia. Type in London (Linn.).

In arid, puffed and wet solonchaks, solonchaklike rubbly desert trails and plains, chee grass thickets, fringes of toirims, banks of saline lakes.

IIA. Junggar: *Jung. Gobi* (between Tungut-Gurban and Khorguto wells, Sept. 22, 1875 — Pias.; Kran river, gravelly descent, Aug. 29, 1876 — Pot.; "Montague pres de l'Ebi-Nor, July 30, 1895, Chaff." — Danguy, l.c.; on Yamatei mountains, Aug. 4; in valley of Chyuizhe springlet, Aug. 8–1898, Klem.; Gashyun-us area, July 1947; Oshigiin-us area nor. or Baitak-Bogdo, granitic hummocky area, rock desert, 30 m, July 31, 1947 — Yun.; on old Guchen road near Erien-Tologoi-ula from south, puffed solonchak, Sept. 13, 1948 — Grub.; Tsitai region, 350 m, hammada - rocky desert — No. 2338, Sept. 26, 1957 — Sheng; same site, 800 m, No. 5178, Sept. 26, 1957 — Kuan; Tsitai region, Beida-Shan'-Dashitou, desert, No. 5262, Sept. 30, 1957 — Kuan; same site, 90 km from Beida-Shan' mountain range, hammada, No. 2439, Sept. 30, 1957 — Shen; left bank of Chern. Irtysh 45 km above Burchum settlement on road to Koktogai, rubbly desert, July 9, 1959 — Yun. et Yuan'; 75 km south-west of Bulgan somon centre, near Oshigiin-us springs, chee grass thicket, Aug. 7, 1977 — Volk. and Rachk.; 80 km south-east of Bulgan somon centre, solonchak, July 29, 1988 — Beket, Vinogradova et al.), *Zaisan* (clay bank of Charynta river, July 5, 1876 — Pot.), *Dzhark.* (on Ili river, June 1, 1877; between Manchurian sumun and Kainak settlement, Aug. 30, 1880 — A. Reg.; 37 km west of Ili town [Kul'dzha] on road to Suidun, Aug. 30, 1957 — Yun. et Yuan'), *Balkh.-Alak.* (Chuguchak basin 10 km south-east of Durbul'dzhin [Emel'] town on road to Toli settlement, chee grass thicket, Aug. 8, 1957 — Yun. et Yuan').

General distribution: Aralo-Casp., Fore Balkh., Jung.-Tarb.; Europe (south. Europ. Russia, Crimea), Fore Asia, Caucasus (east. Trans-Caucasus), Mid. Asia (plains), West. Sib. (south.).

## Section 3. Siphonocalyx Lincz.

6. L. callianthum (Peng) Grub. comb. nova. — *L. drepanostachyum* Ik.-Gal. ssp. *callianthum* Peng in Guihaia, 3, 4 (1983) 292; id. in Fl. Sin. 60, 1 (1987) 40.

—Ic.: Fl. Sin. 60, 1, tab. 6, fig. 7–8.

Described from Siberia. Type in Beijing (PE).

On arid rubbly slopes in wormwood-saltwort deserts.

IB. Kashgar: *West.* (Artux [Artush]—A.-j. Li and J.-n. Zhu, No. 7886, typus !).
General distribution: endemic.

Note. Differs from Mid. Asian *L. drepanostachyum* Ik.-Gal. in much larger flowers
(calyx 6.5–7, not 4.5–6.5, mm long), few-flowered short and rather few spikes and
subglabrous first inner bracts.

## Section 4. Plathymenium Boiss.

7. L. aureum (L.) Hill, Veg. Syst. 12 (1767) 37 ind. tab. 37, fig. 4; Ktze.
Rev. gen. 2 (1891) 395; Kitag. Lin. fl. Mansh. (1939) 354; Lincz. in Fl. SSSR,
18 (1952) 435; Grub. Konsp. fl. MNR [Conspectus of Flora of Mongolian
People's Republic] (1955) 221; Fl. Tsentr. Sib. 2 (1979) 708; Fl. Intramong.
5 (1980) 47; Grub. Opred. rast. Mong. [Key to Plants of Mongolia] (1982)
200; Opred. rast. Tuv. ASSR [Key to Plants of Tuva Autonomous Soviet
Socialist Republic] (1984) 79; Pl. vasc. Helanshan (1986) 196; Peng in Fl.
Sin. 60, 1 (1987) 37, p.p. excl. var. *potaninii* (Ik.-Gal.). Peng.—*Statice aurea*
L. Sp. pl. (1753) 276; Boiss. in DC. Prodr. 12 (1848) 641; Ldb. Fl. Ross. 3
(1849) 458; Turcz. in Bull. Soc. Natur. Moscou, 25, 3 (1852) 896; ?
Henders. et Hume, Lahore to Jarkend (1873) 332; Hance in J. Bot.
(London) 20 (1882) 291; ? Kanitz in Die Resultate...2 (1898) 713; ? Deasy,
In Tibet and Chin. Turkestan (1901) 403; Hemsl. Fl. Tibet (1902) 189; Diels
in Futterer, Durch Asien (1903) 13; Palibin in Tr. Troitskosavsko-
Kyakhtinsk. otdeleniya Priamursk. otdela Russk. Geogr. Obshch. 7, 3
(1904) 49; Gr.-Grzh. Zap. Kitai, 3 (1907) 496; Danguy in Bull. Mus. nat.
hist. natur. 17 (1911) 338; ibid, 20 (1914) 72; Hedin, S. Tibet, 6, 3 (1922)
49; Gr.-Grzh. Zap. Mong. 3 (1930) 820; ? Persson in Bot. notiser (1938)
296; Hao in Bot. Jahrb. 68 (1938) 627; Walker in Contribs U.S. Nat. Herb.
28 (1941) 650; Chen et Chou, Rast. pokrov r. Sulekhe [Vegetational Cover
of Sulekhe River] (1957) 89.—*S. lacostei* Danguy in J. de Bot. 21 (1908)
53.—*S. dielsiana* Wangerin in Feddes Repert. 17 (1921) 399.

—Ic.: Reichb. Pl. crit. 2, tab. 195; Fl. Intramong. 5, tab. 19, fig. 1–4; Fl.
Sin. 60, 1, tab. 6, fig. 1–2.

Described from Siberia (Dauria). Type in London (Linn.).

In solonchaks, banks of saline lakes, solonetzic sand, fringes of toirims
and floors of gorges, in chee grass thickets, desert-steppe rubbly slopes
and mountain trails, in saltwort and saxaul deserts.

IA. Mongolia: *Khobd.* (Kharkhira mountain group, Namyur river, July 18, 1903—Gr.-
Grzh.—extreme nor.-west. find), *Cent. Khalkha, East. Mong.* (in salsis Mongoliae
chinensis 1831—I. Kusnezow; ibid, 1831—Bunge; Mongolia chinensis in itinera ad.
Chinam, 1830 et in reditu e China, 1841—Kirilow; Mongolia chinensis, 1842—Gorski;

Ourato, plaines salcis et humides, No. 2802, July 1860 – David; plain around Kuku-Khoto, in ruins of Tuchen town, Aug. 2; plain nor. or Khekou town on Huang He, Aug. 3–1884, Pot.; Lukh-sume monastery, July 3; Kulun-Buirnor plain, Dzhalatu area, July 8–1899, Pot. et Sold.; lower Kerulen on Dure-Tsagan-nur lake; same site, below Bo-Tszangin-sume; same site, near Bain-Tsagan – 1899, Pal.; Khuna province, around Sinbaerkhuyunchi village, No. 1013, June 29, 1951, and many others throughout the distribution range), *Depr. Lakes* (between Baga-nor and Kirgiz-nor lakes, Aug. 1; between Dzeren-nor lake and Dzankhyn river, Aug. 6–1879, Pot.; lower course of Buyantu river, 1942 – Kondrat'ev; south-west. fringe of Khunsiin-Gobi 5–7 km east of Bain-Gol factory, Aug. 27, 1944 – Yun.) *Gobi-Alt., Val. Lakes, East. Gobi* (many finds in Mongolian People's Republic; vicinity of Bailinmyao town, 1959 – Ivan.), *West. Gobi* (Beishan': Khami desert, July 6, 1879 – Przew.; Shidun, Sept. 10; Baga-Madzhin-syan' Oct. 5–1890, Gr.-Grzh.; south. foothills of Tsagan-Bogdo mountain range, Aug. 1; between Atas-Bogdo and Maikhan-bulak, Aug. 16–1943, Yun.; lowland north of Khabtsagain-Undur-nuru on border road, Aug. 17; 8 km nor.-west of Chonoin-bom area on road to Ekhin-gol, conical hill, 139 m, Aug. 19–1948, Grub.; Beishan', 90 km nor.-east of Tszin't town, July 23, 1958 – Petr.; between Suman-Khairkhan-ula and Ederingiin-nur, July 12, 1973 – Golubkova and Tsogt; south. foothills of Adzhi-Bogdo mountain range 30 km from Altay somon centre on road to Tseel' somon, conical hill, 1450 m, Aug. 24; 48 km from Talyn-Moltys oasis nor.-east of border road, Aug. 26–1979, Grub., Dariima, Muld.), *Alash. Gobi* (in desert, June 7, 1872 – Przew.; Kobden-obotu area, gobi north of Gashiun-nur lake, Aug. 12; Kobden-usu between Gashiun-nur lake and Tostu mountains, Aug. 13–1886, Pot; Tengeri sand, Tszinzi-kho area, 1370 m, Sept. 23, 1901 – Lad.; Dyn'yuan' in oasis, May 31; Baiminte area, June 6–1908, Czet.; "Chung-Wei, No. 230; Holan-Shan, No. 1059, 1075, 1923, Ching" – Walker l.c.; 90 km south-west of Chzhunvei town, June 30; 15 km west of Bayan-Khoto town, somewhat overgrown sand, July 5; vicinity of Chzhunvei town, Tengeri sand, July 28–1957; 90 km south-east of Bayan-Khoto town, Divusumu settlement, south. fringe of Bayan-Nor mountains, June 12; 60 km south of Bayan-Khoto town, east. offshoot of Kholan-shan' mountain range, rocky lowland, June 14; 90 km south of Inchuan' town, Chintunai gorge, on Huang He river, low mountains, June 18; 13 km west-nor.-west of Bayan-Khoto town, south. fringe of Yaburai mountains at Yaburai-yan'chi settlement, June 30; 15 km south-east of Mintsin town, vicinity of Gaotszyasavo town, July 3; south. fringe of Tengeri sand around well and Maachantszin' settlement [near Uvei town], July 23; Dakhuansatan south-west of Bayan Khoto town, Tengeri sand, July 27–1958; 36 km east of Mintsin town, Syaoergou oasis, sand, Aug. 18, 1959 – Petr.), *Ordos* (Huang He river valley, July 21, 1871 – Przew.; Boro-Balgasun area, Sept. 14; Baga-chikyr saline lake, Sept. 25–1884, Pot.; 50 km nor.-west of Khangin-chi town, Yan'khaitszy lake, bank, July 6, 1957 – Petr.), *Khesi* (east of An'si town, Aug. 6, 1875 – Pias.; desert between mountains north of Tetung river and Alashan, Aug. 12, 1880 – Przew.; between Kheichen village and Yan'chi village, June 28, 1886 – Pot.; Suchzhou, July 24, 1890 – Marten; Sachzhou oasis, 1130 m, Aug. 8, 1895 – Rob.; "along lower course of Sulekhc river–Chen et Chou, l.c.; 45 km west of Yunchan town, nor. foothills of Nanshan, July 10; 35 km south-east of Chzhan'e town, nor. trail of Nanshan mountain range, July 14; 55 km west of Tszyutsyuan' town, Aug. 6–1958; 80 km south-east of Dunkhuan, nor. trail of Altyn-tag, Oct. 7, 1959 – Petr.; nor.-east of Tszyutsyuan' town, hummocky area, Sept. 23, 1958 – Lavr.).

IC. Qaidam: *Plain* ("Steinwuste beim Lager 66 am Rande der Tsaidam, district Barun, No. 242, July 14, 1906, A. Tafel", typus *Statice dielsiana* Wangerin, l.c.; 110 km west of Tsagan-Us, gorge along eastern fringe of Qaidam depression, Oct. 13, 1959 – Petr.), *Mount.* (Dulan-khit temple, 3050–3350 m, in rock crevices, Aug. 10, 1901 – Lad.).

IIIA. Qinghai: *Nanshan* (along Kuku-usu river, 2450–2600 m, July 11, 1879 – Przew.; Huang He valley, Yuiguan', Shidun' village, 1400–2450 m, July 1, 1890 – Gr.-Grzh.; Humboldt mountain range, nor. slope along Kuku-usu river, 2750–3050 m, May 12; Yamatyn-umru mountains, 3650–3950 m, July 22; South. Kukunor mountain range, nor.

slope along Noion-Khutul'-gol river, 3950 m, Aug. 3–1894; Humboldt mountain range, nor. slope of Chan-sai gorge, 3350–3650 m, July 22, 1895–Rob.; "in planitie Kulangshien"–Kanitz, l.c.; "Nan-schan bei 2500 m auf der Pass höhe zwischen dem Tatung-Gebiete und dem Sining-Gebiete, bei Ping-kouyi, Futterer"–Diels, l.c.; "Nan-Chan, Che-Yeon-Hu, alt. 2300 m, June 18; Sining-Fou, alt. 2400 m, July 18, 1908, Vaillant"–Danguy, 1911, l.c.; "Kokonor, um den See Ganbadalian-nor, 3150 m, 1930"–Hao, l.c.), *Amdo* ("Thal des Hoangho oberhalb Balekun-gomi, Futterer"–Diels, l.c.; Gunkhai depression 20 km west of Gunkhe town, 2980 m, Aug. 6, 1959–Petr.).

IIIB. Tibet: *Chang Tang* ("in Karakash and Aralan valleys, 9000–12,000 ft" – Henderson et Hume, l.c.; Polu, Ishak-Kurmag, June 5, 1890–Grombch.; nor. slope of Russkoe mountain range, Bashbulak area along upper Aksu river, 3650 m, July 1; same site, near Kara-Sai village, 3050 m, July 8, 1890, Rob.; "Northern Tibet, Mandarlik, 3437 m, July 1900"–Hedin, l.c.; Suget-davan, 1912–Avinov), *Weitzan* (Burkhan-Budda mountain range, on south. slope, 4400–4800 m, Aug. 11; same site, nor. slope along Khatu-gol river, 3100 m, Aug. 23–1884, Przew.; "In 97°–35°42', 13,363 ft. Shuga-gol, Sept. 15, 1896, Wellby and Malkolm"–Hemsl. l.c.; Burkhan-Budda mountain range, Khatu gorge, 3200 m, July 3–1901, Lad.).

IIIC. Pamir (An'elok-daba 10 km below Kizyl-davan, 3000 m, July 6, 1941–Serp.).

General distribution: East. Sib. (south. Tuva, Daur. south.), Nor. Mong. (Mong. Daur.: vicinity of Ulan Bator and Toly river meander), China (Nor.-West.: Gansu).

Note. Varies quite widely. Plants go up to 30 cm height but only 5 cm tall dwarf plants are found in Qinghai and Tibet. Specimens with a cluster of suberect thin stems grow on sand. Inflorescence is very compact, with many-flowered (up to 12 !) as well as very loose, 1–2-flowered spikelets. Flowers in the northern part of the range (in Mongolia) and in Tibet are generally 5–6 mm long while, in the southern part (Qinghai, Ordos, Khesi), the length of flowers goes up to 7–8 mm. Calyx is usually with orbicular lobes but forms with deltoid lobes and up to 0.8 mm long awn are found rather often. Such a form was described from Qaidam as a distinct species, *Statice dielsiana* Wangerin, l.c., but similar forms were noticed along with type in East. Mongolia, Ordos, Khesi, Qinghai. Awn is also seen on orbicular lobes of calyx. Insofar as pubescence of calyx tubes is concerned, it is all over or only along nerves, or tubes subglabrous. Colour of calyx everywhere is invariably bright golden yellow but specimens with infuscated flowers are occasionally found among herbarium specimens. These are predominantly from autumn collections, possibly the result of frosts.

The excellent photographs and twig of inflorescence of type (and unique) *Statice lacostei* Danguy kindly supplied by the Museum of Natural History, Paris (P) helped conclude that it does not differ at all from the common *Limoneum aureum* (L.) Hill = *Statice aureum* L. The feature of flattened branches attributed by Danguy as a distinct characteristic of it is a consequence of flattening of the plant on desiccation while bicoloured spikelets fall within the range of natural variation of *L. aureum* which was already discussed above.

8. L. bicolor (Bge.) Ktze. Rev. gen. 2 (1891) 395; Kitag. Lin. fl. Mansh. (1939) 354; Grub. Konsp. fl. MNR [Conspectus of Flora of Mongolian People's Republic] (1955) 221; Fl. Intramong. 5 (1980) 50; Grub. Opred. rast. Mong. [Key to Plants of Mongolia] (1982) 199; Pl. vasc. Helanshan (1986) 196; Peng in Fl. Sin. 60, 1 (1987) 31.—*Statice bicolor* Bge. Enum. pl. China bor. (1832) 55, No. 303; id. in Mem. Ac. Sci. sav. etrang. 2 (1835) 129, incl. var. α *laxiflora* Bge. and β *densiflora* Bge.; Trautv. in Acta Horti Petrop. 1 (1872) 189; Palib. in Tr. Troitskosavsko-Kyakhtinsk. otdeleniya Priamursk. otdela Russk. Geograf. Obshch. 7, 3 (1904) 50; Danguy in Bull. Mus. nat. hist. natur. 17 (1911) 338; Walker in Contribs U.S. Nat.

Herb. 28 (1941) 650.— *S. bungeana* Boiss. in DC. Prodr. 12 (1848) 642.— *S. varia* Hance in J. Bot. (London) 20 (1882) 290; Kanitz in Die Resultate bot. Samml. 2 (1898) 713.

—Ic.: Fisch. et Mey. Sert. Petrop. (1846) tab. 17; Grub. Opred. rast. Mong. [Key to Plants of Mongolia] Plate 107, fig. 481; Fl. Intramong. 5, tab. 20, fig. 3–7; Fl. Sin. 60, 1, tab. 5, fig. 4.

Described from Nor. China (toward south-east of Kalgan), Mongolia. Syntypes in Paris (P); isosyntypes in St.-Petersburg (LE).

27 In steppes, specially sandy, on steppe rubble slopes of conical hills and low mountains, on solonetzic sand and along sandy floors of gorges, on solonchaks along low lands and banks of saline lakes, in chee grass thickets and rock crevices.

IA. Mongolia: *Khobd.* (east. slope of Dzhindzhidin between Ureg-nur and Ubsa-nur, July 30, 1945 — Yun., extreme north-west. isolated find), *Cent. Khalkha* (plain between Chilotei-bulak and Gil'te-guntui, Aug. 9, 1897 — Klem.; upper Kerulen, Tono-ula mountains, July 22; central Kerulen near Tsagan-Obo mountains, July-Aug.; same site, region of Bare-Khoto ancient township, July-Aug.—1899, Pal.; road to Urgu from Alashan, Sharkhai-khuduk well, July 24, 1909 — Czet.; 80–100 versts-1 verst = 1.067 km-west–south-west of Urga on road to Ikhe-Tukhum-nor, Aug. 8, 1925 — Krasch. et Zam.; near Choiren, July 1, 1926 — Kondrat'ev; Choiren, Bogdo-ula summit, 1683 m, Sept. 2, 1927 — Zam.; Choiren-ula, on rocks, July 7, 1941; Bor-Undur somon, 180 km south-west of Ulan-Bator on Ulyasutaisk road, July 1, 1949 — Yun.; Bogdo-ula mountain near old Choiren, on rocks, July 4, 1970 — Banzragch, Karam. et al; 85 km south–south-west of Ulan Bator along road, smoothed fine hummocky area, Aug. 28, 1972 — Grub., Ulzij. et al; mountains 31 km east–south-east of Dzhargalt-Khan somon centre on road to Undurkhan, 1514 m, July 7, 1979 — Grub., Dariima, Muld.), *East. Mong.* (around Tsagan-Balgas, June 15, 1830 — Ladyzh.; in glareosis inter Tschadao et Tschantsia-keou [Kalgan] Chinese borealis, 1831 — Bge., var. α-syntypus !; in desertis Mongolia meridionalis, 1831 — Bge., var. β-syntypus !; Mongolia chinensis, 1840 — Kirilow; ibid, 1842 — Gorski; Sartehy, plaines deesser. June 1864; Ourato, heutes montaguis dicouvertes, No. 2699, July 1866 — David; Muni-ula, June 28, 1871; Suma-Khada mountains, May 30, 1872 — Przew.; plain south of Guikhuachen [Khukh-Khoto], July 6, 1884 — Pot.; between Ulan-Khada and Tsindai on mall road, Aug. 10, 1898 — Zab.; between Kulun and Buir lakes, north of Dulan-Khara mountains, June 12; south of Bain-nor lake [south. fringe of Buir-nor], June 20*; Ulan-Dzhilgu area, July 3; Dzhalatu locality, July 8; Ergenten-gol river valley, July 10–1899, Pot. et Sold.; lower Kerulen, near Bain-Ul'dzeitu mountains and below Botszangin-sume at the same site — 1899, Pal.; Khailar, July 6, 1901 — Lipsky; near Kharkhonte railway station, Aug. 20, 1902 — Litw.; same site near saline lake, Sept. 10, 1927 — Gordeev; Shavorte-Obo area 55 km east of Erentsab, hummocky area, Aug. 19, 1949 — Yun.[1]; vicinity of Shilin-Khoto town, 1960 — Ivan.), *Gobi-Alt.* (Bayan-Tukhum area 20 km west of Khongor-Obo somon, June 11, 1938 — Luk'yanov), *East. Gobi* (Shara-Murun, hilly grasslands, No. 470, 1925 — Chaney; granitic mountain range 25 km south of Ude workshop, July 19, 1928 — Tug.; Argali mountain range and vicinity of Khodala-khuduk well, Sept. 5, 1928; Ergil'-uda on old road to Ali'-Bayan somon, on precipices, June 30, 1941 — Yun.; Argalantu mountains, nor. slope under Tsagan-Obo summit, Aug. 6, 1970 — Grub., Ulzij. et al), *Alash. Gobi* (Alashan' mountains, nor. portion, July 22, 1873

---

[1]Of several finds in this region within the territory of Mongolian People's Republic, only the specimens from the extreme northern part of distribution range are cited here.

—Przew.; Khoir-Toora area on Edzin-gol river, July 18*; left bank of Edzin-gol river between Khoir-Toora and Gantszy-dzak areas, July 19*—1886, Pot.; Dyn'yuan' in oasis, in shrubs, May 30, 1908—Czet.; "Mouth of Hsi-Yeh-Kou, No. 180 and Holanshan, on dry gravelly slopes, No. 1132, 1923, Ching"—Walker, l.c.; 45 km south of In'chuan' town, first terrace of Huang Ho river, July 3, 1957—Petr.; from In'chuan' to Bayan-Khoto, desert, July 5, 1957—Kabanov), *Ordos* (valley of Huang He river [Aug.] 1871—Przew.; Ordos australis, Autumno 1877—Verlinden; Ushkin'-Tokul' area, Aug. 29; Baga-Chikyr Saline lake, Sept. 26; Nedzhalyan village [on southern boundary of the region], Sept. 28 —1884, Pot.; 15 km west of Ushinchi town, somewhat overgrown sand, Aug. 2; vicinity of Dalatachi town, overgrown sand, Aug. 11; 5 km north of Yuilin' town, sandstone precipices, Aug. 19—1957, Petr.), *Khesi* (Pabatson' valley, slope of loessial mountains, July 24, 1908—Czet.*).

IIIA. Qinghai: *Nanshan* ("Koko-nor"—Hance, l.c.; "in valle Tetung, June 22, 1879" —Kanitz, l.c.; "Nien-Pe, alt. 2200 m, July 11, 1908, Vaillant"—Danguy, l.c.; "Yai-Chieh, No. 249, 1923, Ching"—Walker, l.c.).

General distribution: Nor. Mong. (Mong. Daur.), China (Dunbei, Nor., Nor.-West., Cent.-nor.).

Note. Highly polymorphous species in form and height of stem (up to 60 cm), extent of its branching, abundance of barren branches, compactness and form of inflorescence, and number of flowers in spikelets. Bunge in his description of this species distinguished 2 varieties var. *laxiflora* Bge. with loose, diffuse inflorescence, and var. *densiflora Bge.* with compact, corymbose-paniculate inflorescence. Boissier even elevated the latter to the species rank, *Statice bungeana* Boiss. These are, however, only extreme forms with neither morphological nor geographic boundaries. In fact, the tall var. *laxiflora* is more often found in the eastern and south-eastern parts of the distribution range where the climatic conditions are more favourable. The flowers exhibit the entire gamut of colours from bright raspberry to pure white; calyx tube is usually glabrous but could also be subpubescent. The number of plant stems also varies. Usually, they are unicaulous but plants with 2-3 stems are also seen; multicaulous forms with a full tuft of slender stems and strong root (asterisked, *) are found on solonetzic sand. Typically stems are erect but could be flexuose as well.

Very close to *L. sinense* (Girard.) Ktze. and difficultly distinguished from it, specially var. *laxiflora*.

Evidently, it hybridises with *L. aureum* (L.) Hill. In Huang He valley on Ordos, Przewalsky collected entirely typical samples of *L. bicolor* but with a light yellow calyx (No. 289, July 22, 1871—Przew.). It is possible that the yellow-flowered *L. grubovii* Lincz. is also of hybrid origin.

9. L. chrysocomum (Kar. et Kir.) Ktze. Rev. gen. 2 (1891) 395; Lincz. in Fl. SSSR, 18 (1952) 431; Grub. Konsp. fl. MNR [Conspectus of Flora of Mongolian People's Republic] (1955) 221; Fl. Kazakhst. 7 (1964) 76; Grub. Opred. rast. Mong. [Key to Plants of Mongolia] (1982) 200; Peng in Fl. Sin. 60, 1 (1987) 39, p.p. excl. var.; Lincz. in Opred. rast. Sr. Azii [Key to Plants of Mid. Asia] 10 (1993) 37. — *L. schrenkianum* (Fisch. et Mey.) Ktze. l.c. 396. — *Statice chrysocoma* Kar. et Kir. in Bull. Soc. Natur. Moscou, 15 (1842) 429; Kryl. Fl. Zap. Sib. 9 (1937) 2161. — *S. schrenkiana* Fisch. et Mey. in Bull. Ac. Sci. St.-Petersb. 1 (1843) 362.

—Ic.: Fl. SSSR, 18 Plate 21, fig. 4; Fl. Kazakhst. 7, Plate 8, fig. 11; Fl. Sin. 60, 1, tab. 6, fig. 5–6.

Described from East. Kazakhstan (vicinity of Ayaguza town). Type in St.-Petersburg (LE).

On arid rocky slopes and trails of conical hills and low mountains, on rocks.

IA. Mongolia: *Mong. Alt.* (upper Bulugun river valley, July 24, 1906—Sap. (var. *pubescens* Lincz.), *Depr. Lakes* (nor. fringe of Khirgiz-Nur basin, submontane plain of Khan-Khukhei mountain range, desert steppe, July 17, 1973—Banzragch, Karam. et al).

IIA. Junggar: *Jung. Gobi* (east. extremity of Semistei mountain range 25-30 km from Khobuk settlement on road to Kosh-Tologoi, on rocky, desertified slopes of conical hills, June 24, 1957—Yun., S.-y. Lee, Yuan'; Baitak-Bogdo, nor, foothills, 1900 m, on rocks, Aug. 1, 1988—Gamalei et al).

General distribution: Aralo-Casp., Fore Balkh., Jung.-Tarb.; West. Sib. (south of Irtyshsk., south-west. Altay).

Note. Form with pubescent bracts, var. *pubescens* Lincz., is found in the extreme eastern portion of the distribution range.

10. L. congestum (Ldb.) Ktze. Rev. gen 2 (1891) 395; Lincz. in Fl. SSSR, 18 (1952) 423; Grub. Konsp. fl. MNR [Conspectus of Flora of People's Republic of Mongolia] (1955) 221; id. Opred. rast. Mong. [Key to Plants of Mongolia] (1982) 199; Opred. rast. Tuv. ASSR [Key to Plants of Tuva Autonomous Soviet Socialist Republic] (1984) 79.— *L. iljinii* Sobolewsk. in Not. syst. (Leningrad) 14 (1951) 48.— *Statice congesta* Ldb. Fl. alt. 1 (1829) 437 and Fl. Ross. 3 (1849) 468; Sapozhn. Mong. Altai (1911) 383; Gr.-Grzh. Zap. Mong. 3 (1930) 820.

—Ic.: Ldb. Ic. pl. fl. ross. 4, tab. 314.

Described from West. Siberia (Altay, Chui steppe). Type in St.-Petersburg (LE). Plate I, fig. 1; map 1.

On solonchaks, solonetzic meadows along floors of mountain valleys, in chee grass thickets in mountain steppe belt.

IA. Mongolia: *Khobd.* (Kalgutta river valley [Oigur], June 28, 1905—Sap.; "Oigur, Suok"—Sap. l.c.), *Mong. Alt.* (Iter ad Chobdo, zwischen den 2n und in Piquet, July 12, 1870—Kalning; Saksa river, July 8, 1877—Pot.; Bodonchi river valley, July 19, 1898—Klem.; Tal-Nor lake, 2570 m, July 4, 1903—Gr.-Grzh.; "Kak-Kul', Dain-Gol" - Sap. l.c.; Dzuilin-gol valley 65 km from Tonkhil somon centre on road to Tsastu-Bogdo-ulu, June 23; west. vicinity of Tsetseg-Nur lake, south-west. slope of Temetiin-Khukh-ul on road to Must somon, 2150-2200 m, June 26; Buyantu river basin, Deliun area near Bukhu-Tumur burial ground, July 1—1971, Grub., Dariima, Ulzij.), *Depr. Lakes* (Khara-Usu lake, Aug. 18, 1930—Bar.).

General distribution: West. Sib. (Altay, Chui steppe), East. Sib. (Tuva, Mongun-Taiga).

11. L. dichroanthum (Rupr.) Ik.-Gal. ex Lincz. in Fl. URSS, 18 (1952) 428; Grub. in Bot. mat. (Leningrad) 19 (1959) 548; Lincz. in Opred. rast. Sr Azii [Key to Plants of Mid. Asia] 10 (1993) 37.— *Statice dichroantha* Rupr. in Mém. Ac. Sci. St.-Petersb., 7 ser. 14, 4 (1869) 69.

Described from Cent. Tien Shan (Dzhaman-Daban). Type in St.-Petersburg (LE).

On desert-steppe rocky slopes of mountains.

IB. Kashgar: *East.* (Chichan'-tsun' village to Karangao in Turfan, 1200 m, No. 5767, June 22, 1958—Lee et Chu).

IIA. Junggar: *Tien Shan* (Mertsbakhera mountain range, nor. slope, 1100–1200 m, semidesert mountain plateau in Santai region, June 12, 1952—Mois.).

General distribution: Cent. Tien Shan.

## 12. L. erythrorhizum Ik.-Gal. ex Lincz. in Novit. syst. pl. vasc. 8 (1971) 211; Grub. Konsp. fl. MNR [Conspectus of Flora of Mongolian People's Republic] (1955) 222, nomen; id. Opred. rast. Mong. [Key to Plants of Mongolia] (1982) 200.

Described from Mongolia. Type in St.-Petersburg (LE). Plate I, fig. 2, map 1.

In puffed solonchaks, solonetzes, solonetzic coastal sand, on solonetzic rocky trails and slopes of conical hills, along gorges and fringes of toirims in deserts.

IA. Mongolia: *Gobi-Alt.* (Khurkhu mountain range, Ikhe-Nomogon mountain, Undur-Sudzhi well, July 22, 1930—Simukova; Bayan-Tukhum area, on puffed solonchaks, Aug. 5; same site, puffed solonchak, Aug. 29; same site, west of saline lake, hummocky sand, No. 4553, Sept. 15–1931, Ik.-Gal., typus !; Artsa-Bogdo mountain range, south. trail of mountain range, vicinity of Dzhargalantu well, June 22, 1945—Yun.; same site, nor. slope of Dulan-ula mountain, July 7, 1973—Isach. et Rachk.; Bayan-Tukhum lake basin, nor.-east. bank, 1440 m, puffed solonchak, Sept. 9, 1979—Grub, Dariima, Muld.), *East. Gobi* (Shabarakh-Usu, sandy wash, about 1100 m, flowers pink, No. 93; Gatun-Bologai, rolling uplands at 900 m, No. 445, 1925—Chaney; central Gobi, between Ulan-Sondzhi and Dzak-Obo-Khuduk, July 19, 1926—Glag.; Ail'-Bayan somon, 10–12 km nor.-east of Khoinchin-khrala along telegraphic line, Sept. 10; same site, 15–20 km nor.-east of Khoinchin-khurala on road to Shine-Usu-khuduk, Sept. 19–1940, Yun.; 120 km south-east of Nomogon somon centre, Shiltula, on slope of outlier, July 19, 1974—Rachk, et Volk.), *Alash. Gobi* (between Khara-Obo well and Tsaikhe well [south of Khurkhu mountain range], Aug. 5, 1873; same site, around Khara-Obo well, Sept. 18, 1880—Przew.; road to Urgu from Alashan, Ikhengun area, June 24, 1909—Czet.; desert east of Sogo-Nor lake, Tsyailin-bulak spring, in sand, Sept. 20, 1925—Glag.; Bordzon-Gobi, nor. mountain trail and marginal Khalzan-ula hummocky area, June 18, 1949—Yun.; Nomogon somon, Ulan-Ergen-khid, trail, Sept. 9, 1950—Lavr. et Yun.; 10 km west of Sangin-Dalai somon centre along road, near well, slope of conical hill, July 26, 1970—Grub., Ulzij. et al; near Shuulin post, July 31 and Aug. 1, 1989—Grub., Gub., Dariima).

General distribution: endemic.

## 13. L. flexuosum (L.) Ktze. Rev. gen. 2 (1891) 395; Lincz. in Fl. SSSR, 18 (1952) 420; Grub. Konsp. fl. MNR [Conspectus of Flora of Mongolian People's Republic] (1955) 222; Fl. Tsentr. Sib. 2 (1979) 708; Grub. Opred. rast. Mong. [Key to Plants of Mongolia] (1982) 199; Fl. Intramong. 5 (1986) 60; Peng in Fl. Sin. 60, 1 (1987) 34.—*Statice flexuosa* L. Sp. pl. (1753) 276; Ldb. Fl. Ross. 3 (1849) 462.

—Ic.: Fl. SSSR, 18, Plate 21, fig. 3; Grub. Opred. rast. Mong. [Key to Plants of Mongolia] Plate 107, fig. 483; Fl. Intramong, 5, tab. 20, fig. 1–2; Fl. Sin. 60, 1, tab. 5, fig. 7–8.

Described from East. Siberia. Type in London (Linn.).

On rubbly and rocky steppe slopes and rocks, talus, coastal pebble beds, solonetzic meadows and sand, along gorges.

IA. Mongolia: *Khobd.* (pass from Ureg-Nur basin to Achit-Nur basin, July 30, 1945 —Yun., Turgen mountain range, west. extremity of nor. slope, Gashun area, 12 km below Otogor-Khamar-daba pass on road to Tsagan-nur, July 15, 1971—Grub., Dariima, Ulzij.; Dzelengiin khutul pass south of Ureg-Nur, July 17, 1971—Karam., Sanczir et al), *Mong. Alt.* (Saksa river, July 8, 1877—Pot.; nor. trail of Khan-Taishiri mountain range 10 km south-east of Yusun-bulak, July 14, 1947—Yun.; Tamchi-daba pass, about 2700 m, Sept. 7, 1948—Grub.; Dzuilin-Gola valley 5 km from somon centre on Tsastu-Bogdo-ul road, June 23, 1971—Grub., Dariima, Ulzij.), *Cent. Khalkha* (on mountain between Nalaikha and Gagtsa-khuduk, Sept. 1830—Kirilow; in Borokhchin area, June 23; on bank of Kharukhi river, June 25—1895, Klem.; basin of Dzhargalante river, between Botoga and Agit mountains; July 11; same site, between Ubur-Dzhargalante sources and Agit mountain, Aug. 11; same site, Ubur-Dzhargalante valley near Ubur-Duryn river, Aug. 10; same site, waterdivide of Ara- and Ubur-Dzhargalante, Aug. 12; same site, sources of Kharukhe river, Ulan-Khada mountain, Sept. 18–1925, Krasch. et Zam.; vicinity of Ikhe-Tukhum-nur, Temeni-ama gorge, June 1926—Zam.; Sorgol-Khairkhan ridge 70 km south of Ulan-Batora, on road to Dalan-Dzadagad, lowland between conical hills, July 13, 1950—Kal.), *East. Mong.* (vicinity of Manchuria railway station, May 6, 1908—Kom.; same site, 1915—Nechaeva; "Kulun-Buir. region, west."—Fl. Intramong. l.c.), *Gobi-Altay* (Dundu-Saikhan, midbelt, July 9, 1909—Czet.; Dzun-Saikhan, July 19, 1930—Simukova; Dundu-Saikhan, Aug. 17; same site, on rocks, Aug. 19–1931, Ik.-Gal.; Dzun-Saikhan, central and lower belts, June 12, 1945—Yun.; Dundu-Saikhan, south slope, 35 km south-west of Dalan-Dzadagad, July 20, 1950—Kal.; Dzun-Saikhan, nor. slope near pass in Ëlyn-ama gorge, about 2400 m, July 21; same site, Ëlyn-ama gorge 1 km below camping site, right lateral creek valley, July 22–1970, Grub., Ulzij. et al; Dundu-Saikhan, south-east. part of nor. slope, July 12, 1974—Rachk. et Volk.).

General distribution: East. Sib. (vicinity of Irkutsk, Daur.), Nor. Mong. (excluding Cis-Hing.).

Note. Varies widely in plant height, from 5 to 50 cm.

14. **L. gobicum** Ik.-Gal. in Acta Inst. bot. Ac. Sci. URSS, ser. 1, 2 (1936) 260; Grub. Opred. rast. Mong. [Key to Plants of Mongolia] (1982) 200.

Described from Mongolia. Type in St.-Petersburg (LE).

In solonchak-like chee grass thickets.

IA. Mongolia: *East. Gobi* (road from Alashan to Urgu; on road to Tsatego-Tsagan well from Bulygin-Urto area, derris bushes, No. 78, June 6, 1909—Czet., typus !).

General distribution: endemic.

15. **L. grubovii** Lincz. in Bot. zh. 56, 11 (1971) 1635 and 1642; Grub. Opred. rast. Mong. [Key to Plants of Mongolia] (1982) 199.

Described from Mongolia. Type in St.-Petersburg (LE)

In solonetzic arid steppes and coastal solonchak-like meadows.

IA. Mongolia: *East. Mong.* (Tamtsagskii ledge, Lag-Nur lake, south. fringe of lake basin, solonetzic snakeweed-wild rye steppe, Aug. 13, 1970—Grub., Ulzij., Tserenbalzhid, typus !).

General distribution: endemic.

32

16. L. hoeltzeri (Rgl.) Ik.-Gal. ex Lincz. in Fl. URSS, 18 (1952) 426; Fl. Kirgiz. 8 (1959) 167; Lincz. in Opred. rast. Sr. Azii [Key to Plants of Mid. Asia] 10 (1993) 36. — *L. amblyolobum* Ik.-Gal. in Acta Inst. bot. Ac. Sci. URSS, ser 1, 2 (1936) 370. — *Statice hoeltzeri* Rgl. in Acta Horti Petrop. 5 (1877) 259. — *S. tenella* auct. non Turcz,: Rgl. in Acta Horti Petrop. 6, 2 (1880) 385, p.p.

— Ic.: Ik.-Gal. l.c. (1936) fig. 6.

Described from Cent. Tien Shan (upper Chu river). Type in St.-Petersburg (LE)

On rocky slopes of mountains and plateau.

IIA. Junggar: *Tien Shan* (südliches Kiukoñik Tal, June 15, 1908 — Merzb.; ? 32 km east of Ulugchat settlement on road to Kensu mine, badland, June 18, 1959 — Yun. and Yuan', unpub.).

General distribution: Cent. Tien Shan.

Note. Differs quite well externally from *L. kaschgaricum* (Rupr.) Ik.-Gal. with large multicipital caudex up to 8 cm in diam. and many low thickish stems.

17. L. kaschgaricum (Rupr.) Ik.-Gal. in Acta Inst. bot. Ac. Sci. URSS, ser. 1, 2 (1936) 255, in observ.; Lincz. in Fl. SSSR, 18 (1952) 424; Fl. Sin. 60, 1 (1987) 35, excl. syn.; Lincz. in Opred. rast. Sr. Azii [Key to Plants of Mid Asia] 10 (1993) 36. — *Statice kaschgarica* Rupr. in Mem Ac. Sci. St.-Petersb. 7 ser. 14, 4 (1869) 69. — *S. tenella* auct. non Turcz.: Danguy in Bull. Mus. nat. hist. natur. 17 (1911) 339.

— Ic.: Fl. Sin. 60, 1, tab. 5, fig. 9.

Described from Sinkiang. Type in St.-Petersburg (LE).

Along mountain trails and slopes, badlands and in low mountains, along gorges in rocky, predominantly saltwort deserts.

IC. Kashgar: *Nor.* (Ui-tal river gorge at foot of mountains, June 2; south. foothills of Tien Shan, 1650–2000 m, June 3; Kara-Teke mountains, valley of Taushkan-dar'ya river, June 5–1889, Rob.; vor Abad, ausgang des südlichen Musart Tales, May 30; Dschart Tal, steppe vor dem Taleingang, June 14–17; bei Schaichle und Outatür, June 1/2; hinten im Kum-Aryk Tal, auch vor der Eingang zum Dschanart Tal, sehr viel, Mitte June; Oberstes Dschanart Tal bis hinauf zum Pass, July 14–17–1903, Merzb.; "Province de Koutchar, Zamuschtag, July 30, 1907, Vaillant"-Danguy, l.c.; vicinity of Uch-Turfan, barren gorge of Kukurtuk, May 27, 1908 — Divn.; between Aksu and Kucha near Dzhurga village, low arid hillocks, Aug. 13, 1929 — Pop.; west. extremity of Baisk basin 58 km from Bai settlement on highway to Aksu, barren badland, Sept. 21, 1957 — Yun., S.-y. Lee. Yuan'; 25–30 km nor.-west of Kucha town on road to Bai, badland, Aug. 31; nor.-east. part of Bai basin 23 km nor.-east of Kyzyl settlement on road to Kein, badland, Sept. 4; right flank of Taushkan-dar'ya river valley 30–35 km south-west of Uch-Turfan, rocky slopes, Sept. 17; Kurukurzhum-Gobi desert south-east of Uch-Turfan 10 km south of Ak-yar settlement, along trail of ridge and same site, hammada — rocky desert — along badland, Sept. 17, 1958 — Yun. et Yuan'; west of Kucha, in pass, 1350 m, No. 8122, Aug. 31; 4 km nor. of Bain-kiko in Uch-Turfan, 1710 m, No. 8423, Sept. 18–1958; Kucha region, 2020 m, rocky slope, No. 9951, July 24, 1959 — Lee et Chu), *West.* (In regione sylvatica jugi Thian-Schan, Tayun-thal, July 30, 1867 — Osten-Sacken, typus !; 17 versts — 1 verst = 1.067 km — from Irkeshtam, vicinity of mountains, Aug. 11, 1913 — Knorring; Egina to Sim-khake,

July 1; Kizyl-su basin, Ulugchat town, in red sandstones, July 2–1929, Pop.; Kel'pin-
32 bulak in Artush, 1300 m, No. 7490, Sept. 10, 1958—Lee et Chu; 32 km east of Ulugchat
settlement on road to Kensu mine, badland, June 18; 10–12 km nor. of Baikurt settlement
on road to Torugart, 2400 m, steppified desert along rocky slope, June 20–1959, Yun. et
Yuan'; 25 km from Ulugchat along road, Baikurt settlement, 3000 m, on slope, No. 9710,
June 20, 1959—Lee et Chu).

General distribution: Cent. Tien Shan and Issyk-kul basin desert; Mid. Asia (Trans-
Alay mountain range).

18. L. klementzii Ik.-Gal. in Acta Inst. bot. Ac. Sci. URSS, ser. 1, 2
(1936) 361; Grub. Konsp. fl. MNR [Conspectus of Flora of Mongolian
People's Republic] (1955) 222; id. Opred. rast. Mong [Key to Plants of
Mongolia] (1982) 200.

Described from Mongolia. Type in St.-Petersburg (LE). Plate I, fig. 3,
map 3.

In sandy-rubbly deserts and barren steppes, on solonetzic coastal sand
and pebble beds.

IA. Mongolia: *Mong. Altay* (between Tunkul' lake and Kharatei mountains, beyond
which lies Khulmu lake, Aug. 2, 1896—Klem., typus !; valley of Dundu-Tsenkhr river
[Tuguryuk], Aug. 15, 1930—Bar.; 6 km south of Tonkhil-Nur lake on road to Tamchi
somon, on saddle structure, July 16; 12 km nor. of Tamchi somon, peneplanated granitic
hummocky area, July 16–1947, Yun.; south-west. fringe of Tsetseg-Nur basin, gently
inclined mountain trial, June 25; west. fringe of Tonkhil-Nur basin south of Kholbo-
Ulan-ula, along trail of hummocky area, June 26; valley of Tuguryuk-gol 10 km from
Mankhan somon on left-bank road, 1480 m, June 27; Botkhon-gol gorge 13 km from
estuary, 1730 m, June 27–1971, Grub., Dariima, Ulzij.; Tsetseg-Nur basin west of Tsetseg
somon centre on road to Must somon, rocky trail, Aug. 11, 1979—Grub., Dariima,
Muld.), *Depr. Lakes* (Baga-Nor in basin of Khirgiz-Nur, conglomerate mounds, July 31,
1879—Pot.; Tuguryuk-gol area, Aug. 29; near Tymetin-khuduk well, Sept. 7–1899, Lad.;
solonchaks south of khara-Usu lake, Aug. 16, 1930—Bar.).

General distribution: endemic.

19. L. leptolobum (Rgl.) Ktze. Rev. gen. 2 (1891) 395; Lincz. in Fl.
SSSR, 18 (1952) 429; Fl. Kazakhst. 7 (1964) 74; Peng in Fl. Sin. 61, 1 (1987)
34; Lincz. in Opred. rast. Sr. Azii [Key to Plants of Mid. Asia] 10 (1993)
36.—*Statice leptoloba* Rgl. in Acta Horti Petrop. 6, 2 (1880) 385.—? *S. tenella*
auct. non Turcz.: Danguy in Bull. Mus. nat. hist. natur. 20 (1914) 73.

—Ic.: Gartonfl. 30, tab. 1045; Fl. Sin. 60, 1, tab. 5, fig. 6; Fl. Kazakhst.
7, Plate 8, fig. 9.

Described from Sinkiang. Type in St.-Petersburg (LE).

On solonetzic sand and sandy and rubbly trails, in saltwort deserts
and saxaul groves.

IIIA. Junggar: *Tien Shan* (Kul'dzha: Bayandai near Kul'dzha, 900 m, July 11, 1877;
mouth of Taldy gorge, 900–1200 m, May 14; same site, 900–1030 m, May 15, 1879—A.
Reg.; "montague entre le Sairam-Nor et l'Ebi-Nor, July 28, 1895, Chaff."—Danguy, l.c.),
*Jung. Gobi* (lower Borotala, 450–600 m, Aug. 22; vicinity of Takianzi, 300 m, Aug. 24;
südöstl. v. Schicho, 3000–5000 ft [south-east. of Shikho, 900–1500 m], Aug. 25–1878, A.

Reg., typus !; south. fringe of Dzosotyn-Elisun sand on right bank of Manas river 10–15 km north of 21st regiment state farm, saxaul thicket on contaminated sand, June 11, 1957 —Yun., S.-y. Lee, Yuan'; Savan area, 3 km nor.-west of San'daokhotsza, No. 28, June 11; 13 km south-west of Paotai settlement, on roadside and between dunes, No. 77, June 12; 3 km south-west of Paotai settlement, same site, No. 79, June 12; on Paotai-Shamyn'tsza road nor.-west of Paotai, on roadside, No. 152, June 17–1957, Shen; right bank of Manas river, 20–25 km nor. of Damyao on road to Mo-Savan, saxaul grove along sand ridges, July 10, 1957—Yun., S.-y. Lee, Yuan'; 15 km nor. of Usu, No. 2182, Sept. 2; 34 km east of Urumchi, 530 m, slopes of mounds, No. 1847, Sept. 17; ? No. 181, June 20; ? No. 486, July 9; 40 km nor. of Gan'khetsza, No. 2278, Sept. 23–1957, Shen; 33 km east of Shikho, wormwood-saltwort barren land, Aug. 30; 55 km west of Urumchi, 600 m, forbs-saltwort barren land, Oct. 2–1959, Petr.).

General distribution: Fore Balkh. (valley of upper Ili and Charyn river).

20. **L. potaninii** Ik.-Gal. in Acta Inst. bot. Ac. Sci. URSS, ser. 1, 2 (1936) 256. — *L. aureum* (L.) Hill var. *potaninii* (Ik.-Gal.) Peng in Fl. Sin. 60, 1 (1987) 38.

—Ic.: Ik.-Gal. l.c. fig. 2.

Described from Qinghai. Type in St.-Petersburg (LE). Map 3.

On rocks and arid rocky mountain slopes, in wormwood-cereal grass-saltwort semideserts.

IA. Mongolia: *Khesi* (vicinity of Lanzhou town, Baishan' mountains, on steep slopes, June 24, 1957—Kabanov; Lanzhou, Uchan-Shan' mountain south of town, arid slopes, June 25; 67 km north of Lanzhou town, wormwood-cereal grass-saltwort semidesert, June 29–1957, Petr.).

IIIA. Qinghai: *Nanshan* (at Ti-tszya-pu village between Lanzhou and Pinfan, July 11, 1875—Pias.; along Itel'-gol river, June-Aug. 1885—Pot.; nor. bank of Kuku-Nor, Huang He river, July 1, 1890—Gr.-Grzh.; San-shan-tan village, along rocks, July 22, 1908—Czet.; typus !; Minkhe, 100 km east of Xining town, on road, rocky slopes, Aug. 3, 1959—Petr.).

General distribution: endemic.

21. **L. roborowskii** Ik.-Gal. in Acta Inst. bot. Acad. Sci. URSS, ser. 1, 2 (1936) 255; Fl. Sin. 60, 1 (1987) 37.

—Ic.: Ik.-Gal. l.c. fig. 1; Fl. Sin. 60, 1, tab. 6, fig. 3–4.

Described from Sinkiang. Type in St.-Petersburg (LE). Map 3.

In subshrubby, mostly saltwort, rocky deserts, along gorges and washed out slopes, on rocks.

IB. Kashgar: *Nor.* (south. slope of Kara-Teke mountains, 6000–8000 ft, shale rocks, June 9, 1889—Rob., typus !; Uch-Turfan, vicinity of surgun (?), rocky desert, May 16, 1908 —Divn.; Kurkurzhum-Gobi desert south-east of Uch-Turfan oasis, 10 km south of Ak-yar settlement, hammada—rocky desert—along badland, in gorge, Sept. 17, 1958—Yun. et Yuan'; Aksu, 4 km nor. of Takarichi, 1300 m, in pebbly gobi, No. 8417, Sept. 17, 1958—Lee et Chu), West. (upper Kizil-su river near Shur-bulak village, in red sandstone rocks, July 4, 1929—Pop.; 25 km west of Upal village, No. 562, June 9, 1959—Lee et Chu; nor. trails of King-tau mountain range 25 km south-west of Upal oasis, *Ephedra-Reaumuria* desert, along gorges, June 9; valley of Gëz-dary'ya river 40–42 km south of Upal oasis on road to Tashkurgan from Kashgar, *Sympegma* desert, June 15; 42 km east-nor.-east of Artush oasis on road to Aksu from Kashgar, trails of Kelpin mountain range, hammada—rocky desert—June 21, 1959—Yun. et Yuan'), *South.* (Keriya river basin, 70 km south—south-west of Keriya oasis on road to Polur, 2200 m, foothills, along slopes of ravines, May 10; 10 km nor. of Polur settlement, on road to Keriya, *Sympegma* desert belt, along south. slopes, May 13–1959, Yun. et Yuan').

General distribution: endemic.

22. **L. semenovii (Herd.) Ktze.** Rev. gen. 2 (1891) 396; Lincz. in Fl. SSSR, 18 (1952) 432; Grub. Konsp. fl. MNR [Conspectus of Flora of Mongolian People's Republic] (1955) 222; id. in Bot. mat. (Leningrad) 19 (1959) 548; Fl. Kazakhst. 7 (1964) 78; Grub. Opred. rast.. Mong. [Key to Plants of Mongolia] (1982) 200; Lincz. in Opred. rast. Sr. Azii [Key to Plants of Mid. Asia] 10 (1993) 37.— *L. chrysocephalum* (Rgl.) Lincz. in Fl. URSS, 18 (1952) 434.— *L. sedoides* (Rgl.) Ktze. Rev. gen. 2 (1891) 396; Lincz. in Fl. SSSR, 18 (1952) 434.— *L. chrysocomum* var. *semenovii* Peng, var. *chrysocephalum* Peng and var. *sedoides* Peng in Fl. Sin. 60, 1 (1987) 39. — *Statice semenovii* Herd. in Bull. Soc. Natur. Moscou, 41 (1868) 398.— *S. chrysocephala* Rgl. in Acta Horti Petrop. 6, 2 (1880) 383 and 384 in nota; Gr.-Grah. Zap. Kitai, 3 (1907) 496.— *S. sedoides* Rgl. l.c.: 384 and in nota.

—Ic.: Fl. Kazakhst. 7, Plate 8, fig. 12.

34     Described from East. Kazakhstan (Katu-tau mountains). Type lost; neotype (Turaigyr mountain range) in St.-Petersburg (LE).

In desert-steppe rocky slopes and trails of conical hills and low mountains, on solonetzic sand, saxaul groves, along cliffs and talus, specially on Tertiary variegated rocks.

IA. Mongolia: *Mong. Alt.* (Tsagan-Dugyu, on sandy soil, July 11, 1877—Pot.; Khasagtu-Khairkhan mountain range, west. slope of Shilust-Obo ula, 2150 m, Sept. 13, 1982—Beket et Buyan-Orshikh), *Val. Lakes* (Daying-gol, dry hillsides at 4300 ft., No. 197, 1925—Chaney; 50 km south of Bayan-Khongor town, hummocky region, June 23, 1974—Rachk. et Volk.).

IB. Kashgar: *Nor.* (Khaidyk-gol river, tributary of Tsagan-Usu-gol, 1650 m, steppe, Aug. 12, 1893—Rob.; mountains near Ishma village, between Kucha and Kurlya, Aug. 22; between Kuruk-tag and Chol-tag at Sokur village, on arid ridges, Aug. 28–1929, Pop.; 30 km nor.-west of Khetszin settlement at Bagrashkul' lake, 1600 m, No. 1810, Sept. 1, 1957—Shen), *East.* (south. Tien Shan foothills [south of Nanshan-kou pass], on rocky soil, May 27, 1877—Pot.).

IIA. Junggar: *Jung. Alat.* (südabhang des dschungarischen Alatau, obere Borotala, 6000 ft [1850 m], Aug. 7, 1878—A. Reg., typus *Statice sedoides* Rgl.; seitental des oberen Manass [Chustai] Tales, July 23–25, 1908—Merzb.; arid Tuz-agny ravine [Karamai], uninundated terrace 10 km beyond oasis, June 14, 1954—Mois.; 10 km from Otu village, between dunes, No. 871, Aug. 3, 1957—Shen; 8 km nor. of Borotala [Bole] on Ili-Urumchi road, No. 4767, Aug. 29, 1957—Kuan; 12 km west of Ven'tsyuan', No. 4567, Oct. 21, 1958—Kuan**). *Tien Shan* (gorge mouth of Taldy river, 900 m, May 14, 1879—A. Reg.; Aryslyn, 2750 m, July 11*; Algoi, 1850–2450 m, Sept. 12–1879, A. Reg.*; Boro-Khoro mountain range, between Dzhus-agach and mountains, No. 11, 1879—Gr.-Grzh.; vicinity of Urumchi town, Khunmiodza upland, rocky summits, Sept. 10, 1929—Pop; nor. slope of Merzbacher mountain range 25 km south-west of Santai, Khuansan and Udungou interfluvive region, about 1300 m, June 11, 1952—Mois; pass 93–95 km east of Muleikhe on road to Khami from Guchen, rocky slope on saddle structure, desert-steppe belt, Oct. 5, 1957—Yun., S.-y. Lee, Yuan'*; on Bogdo-Shan in Turfan, 1800 m, No. 5756,

June 19, 1958—Lee et Chu*; 20 km north of Sairam-Nur lake, 1500 m, intermontane plain, wormwood-cereal grass steppe, Aug. 31, 1959—Petr.**), *Jung. Gobi* (west of Turkyul' lake, rocky steppe, June 15, 1877—Pot.*; along Urtaksary river, tributary of Borotal, 1200–2150 m, Aug. 20*; south-east of Shikho, 900–1500 m, July 25; Borborogusun gorge, 900 1200 m, Aug. 23*—1878; Borboro-gusun, 3000–4000 ft [900–1200 m], April 28, 1879—A. Reg., typus *Statice chrysocephala* Rgl.; Bainamun at Dzhina, 1500–1850 m, June 5; Tsagan-usu, Dzhina branch, 975 m, June 8*–1879, A. Reg.; 2 km beyond bridge on Yantszykhai river [Bain-gol] on Manas-Shikho highway, ridge surrounding site of river breaching it, badland, July 7, 1957—Yun., S.-y. Lee, Yuan'; same site, ? No. 462, July 7, 1957—Shen; Ebi-Nur lake basin 10–12 km north of Utai settlement on road to Borotal, saxaul hammada—rocky desert-along trails of hummocky region, Aug. 12, 1957—Yun., S.-y. Lee, Yuan'; 10 km south of Bole, 720 m, No. 2170, Aug. 29; same site, No. 2171, Aug. 29*–1957, Shen; 9 km north of Kosh-Tologoi settlement on Khobuk river on highway to Altay settlement from Karamai, desert steppe along hummocky area, July 4, 1959—Yun. et Yuan'*).

General distribution: Fore Balkh. (far east), Jung.-Tarb. (Jung. gateway).

Note. *L. semenovii* var. *chrysocephalum* (Rgl.) Grub. in Novon, 4 (1994) 31 is marked with a single asterisk (*) and *L. semenovii* var. *sedoides* (Rgl.) Grub. ibid, with 2 (**). Both these varieties are related to the main var. *semenovii* through intermediate forms and are found sporadically in different parts of distribution range of the species, their distinctive features evidently being related only to the actual habitat conditions. Moreover, as in the case of *L. chrysocomum* (Kar. et Kir.) Ktze., in the western part of the range, plants are found not with pubescent (typical of this species) but with capitate bracts (var. *glabra* Lincz.) which is possibly a consequence of hybridization between these species in the zone of overlapping of their distribution ranges.

23. **L. tenellum** (Turcz.) Ktze. Rev. gen. 2 (1891) 396; Grub. Konsp. fl. MNR [Conspectus of Flora of Mongolian People's Republic] (1955) 222; Hanelt et Davazamc in Feddes Repert. 70, 1–3 (1965) 50; Fl. Intramong. 5 (1980) 49; Grub. Opred. rast. Mong. [Key to Plants of Mongolia] (1982) 200; Pl. vasc. Helanshan (1986) 196; Fl. Sin. 60, 1 (1987) 35.—*Statice tenella* Turcz. in Bull. Soc. Natur. Moscou, 5 (1832) 203; Boiss. in DC. Prodr. 12 (1848) 641; Palibin in Tr. Troitskosavsko-Kyakhtinsk. otdelen. Priamursk. otdela Russk. Geogr. Obshch. 7, 3 (1904) 50.

—Ic.: Grub. Opred rast. Mong. [Key to Plants of Mongolia] Plate 107, fig. 482; Fl. Intramong. 5, tab. 19, fig. 5–6.

Described from Mongolia. Type in St.-Petersburg (LE). Map 5.

In rocky and rubbly submontane plains, ridges, slopes and trails of conical hills and low mountains, on rocks and talus, along gorges, on solonetzic sand and in chee grass thickets.

IA. **Mongolia**: *Val. Lakes* (Tui river between Ulan-erge area and Udzhyum, Sept. 6, 1886—Pot.; on mountains on left bank of Tuingol, July 9; same site, on right bank, July 9; in gobi north-east of Orok-nur lake, July 14–1893; between Uta and Baidarik rivers, June 19, 1894—Klem.; barren mountains along Tuin-gol river, Sept. 2, 1924—Pavl.; Ongiin-gol near Lamain-khure, slopes of Khailyaste gorge, Aug. 21, 1926—Lis.; lower course of Tuin-gol river, north-west. border of Durbul'dzhin lake, Oct. 25, 1940—Yun.; Interfluvine zone of Tatsiin-gol and Tuin-gol in Abzag-ula area and Adagiin-Kharakhuduk, rocky slope of Lugaryn-ama canyon, June 13, 1971—Grub., Dariima, Ulzij.; 12 km west—south-west of Teg mine, intermontane plain, June 25, 1972—Banzragch, Karam. et al), *Gobi Alt.* (Tostu mountains, Udzhyum valley, Aug. 16; pass through

Tostu mountain range, Aug. 18; nor. fringe of Bain-Tsagan area, Aug. 26–1886, Pot.; Dundu-Saikhan mountains, south. valley, July 2, 1909 – Czet.; south. portion of Dulema conical hill, plain, Aug. 7, 1924 – Pakhomov; nor. foothills of Ikhe-Bogdo mountain range, May 27, 1926 – E. Kozlova; Ikhe-Bogdo mountain range, foothills and lower belt, Aug. 18, 1926 – Tug.; Khurkhu mountain range, Ikhe-Nomogon-ula, July 22, 1930 – Simukova; Bayan-Tukhum area, Aug. 4; Dzolen mountain range, Sept. 3; foothills of Dzolen mountain range, Sept. 8–1931, Ik.-Gal.; Bayan-Tsagan mountains at Ulan-khuduk well, Oct. 14, 1931 – Krupenin; Gurban-Saikhan mountains, Khurmein somon, June 3, 1939 – Surmazhab; Noyan-Bogdo mountains 1–2 km south of Noyan somon centre, crest of conical hill, July 28, 1943 – Yun.; Artsa-Bogdo mountain range, east. extremity, south. spur of Khaldzan-Khairkhan-ula, July 19; same site, south. trail along road to Leg somon 5 km west of Dzhargalant-khuduk, July 22; hummocky area south of Ikhe-Bayan-ula in Tavun-Khobur-khuduk well, alt. 1703 m. Aug. 2; Nomegetu-nuru mountain range, west. extremity, Khara-Obo summit, Aug. 7–1948, Grub.; "Schotterflache am Nordfluss des Ich-Bogd, No. 2747, June 1962" – Hanelt et Davazamc, l.c.; Barun-Saikhan mountain range, west. slope of hummocky area, June 25, 1970 – Sanczir; Ikhe-Nomogon mountain range, nor. slope, gorge under main summit along trail, July 25, 1970 – Grub., Ulzij., Tserenbalzhid; Khuren-Khana-nuru mountains 40 km nor. of Obotu, in gorge, July 25, 1972 – Rachk. et Guricheva; Tostu-ula, Demiin-usu well, slopes of conical hill, July 14, 1973 – Golubk. et Tsogt; Tostu mountain range, nor. trail 6 km east of Ulan-Tologoi-khuduk well on road to Gurban-Tes somon, 2200 m, Sept. 4; Khuren-Khana-nuru mountain range, Musaryn-Khundei gorge at its entrance, 1800 m, trail, Sept. 7–1979, Grub., Dariima, Muld.), *East Gobi* (in montosis lapidosis Mongoliae chinensis, 1831, I Kuznetsov, typus !; Mongolia chinensis, 1831 – Bunge; vicinity of Ude station, Aug. 2, 1831 – Ladyzh.; Mongolia chinensis in reditu e China, 1841 – Kirilow; Mongolia, Ude, June 5, 1850 – Tatarinov; Sair-usu along mail route, Aug. 3, 1898 – Zab.; Shabarakh-Usu, upland plains at 4100–4600 ft., No. 517, 1925 – Chaney; Baga-Ude, at Urgun-Ulan-khuduk well, Aug. 15, 1926 – Bulle; Baga-Ude, Khara-ula, ravine, Aug. 16; Sain-usu, hummocky sand, Aug. 20–1926, Lis.; Ude station along Kalgan road, July 17; 25 km south of Ude station, granitic mountain range, July 19–1928, Tug.; Del'ger-Khangai somon, Khoir-Ul'dzeitu-Sharangad area, Sept. 14, 1930 – V. Kuznetsov; Del'ger-Khangai mountains, along slope, July 30 and Aug. 1931 – Ik.-Gal.; vicinity of Baga-Ude well, on Bayan-Gote mountains, Aug. 21; 17 km north of Dzamyn-Ude, Khukh-Tologoi well, along slope of same-named mountains, Aug. 28; 40 km north of Dzamyn-Ude, Motonge mountains, Aug. 30–1931, Pob.; Del'ger-Khangai mountain range, slopes, July 1933 – Simukova; Ail'-Bayan somon, 2 km west of Ulegei-Khida, steep slopes of conical hill, Sept. 21, 1940, 50 km south-west of Sain-Shanda on road to Khubsugul somon, plain, June 9; Erdzni somon, nor. fringe of Borokha–gala area, desert steppe, June 17–1941; Kholtu somon, Ushiin-khundei area, June 15–16, 1945; Toli-Obo area 30–40 km from Undur-Shila somon centre, slopes of conical hills, July 27, 1946; Gurban-Saikhan somon, Sumbur-ula east of Tabyin-chzhisa, June 23, 1949 – Yun.; 6 km north-east of Oldakhu-khit monastery ruins, plain, July 15, 1950 – Kal.; plateau north of Gurban-Saikhan, between Talain-bulak and Nyutsugtu-khuduk, Sept. 7, 1950 – Kal., Lavr., Yun.; Byailinmyao town, desert steppe, 1960 – Ivan.; 117 km south – south-west of Mandal-Gobi town on road to Dalan-Dzadagad, plain, July 6, 1970 – Banzragch., Karam. et al; 16 km south of Altan-Shire [somon centre], July 7; 20 km south-west of Saikhan-Dulan somon centre, Buralyn-ula, 1328 m, on summit, July 18; 90 km south of Khubsugul somon centre, July 30–1971, Rachk. et Isach., plain 7 km south – south-west of Bulgan somon centre, July 25, 1975 – Kazantseva), *West. Gobi* (1 km nor. of Burkhantu-bulak spring [Dzun-Mod], along gorge floor, Aug. 23, 1948 – Grub.; "Felsböden am Rand der oase Dzunmod, No. 1010, June 1962" – Hanelt et Davazamc, l.c.), *Alash. Gobi* (Kobdon-usu area north of Gashyun-nor, Aug. 14, 1886 – Pot.; Shardyn [Shartszan]-sume temple, in granite crevices, May 12, 1909 – Czet.; between Khara-Morite and Burgastai, May 15; from Kharmykte-bulak spring to Unyugyute area, June 2 – Napalkov; 2 km from Noyan somon centre, hummocky area, July 25; 30–32 km west—south-west of Noyan somon on road to Obotu, hummocky area, July 26–1943, Yun.; 35 km south-west of Inchuan' town, submontane plain, July 5; 75 km south of Inchuan' town, plain, July 18–1957; 35 km south-west of Inchuan' town, plain, June 10; 25 km south of Bayan-Khoto town, submontane plain, June 10; Bayan-Nor mountains 70 km south east of Divusumu village [Boyan-Khoto district], June 11–1958, Petr.), *Ordos* (40 km south of Denkou town, Alabusu-shan' mountains, intermontane valley, June 8, 1958—Petr.).

General distribution: endemic.

## Family 94. OLEACEAE Hoffmgg. et Link

1.  Leaves pinnately compound. Fruit—lanceolate samara ...............
    ................................................................... 1. Fraxinus L.
+   Leaves entire or pinnatisected ..................................................... 2.
2.  Fruit—bivalved capsule ........................................... 2. Syringa L.
+   Fruit—binary or simple dry globose drupe ........ 3. Jasminum L.

## 1. Fraxinus L.
### Sp. pl. (1753) 1057

1.  Leaflets large, 2.5–8 cm long, 1.5-4 cm broad, on 0.5–1.2 cm long petiolules, with 8–12 pairs of lateral nerves ...................................
    ................................................................ 1. F. sogdiana Bge.
+   Leaflets small, 3–4 cm long, 0.5–1.5 cm broad, subsessile, with 4 pairs of lateral nerves ......... 2. F. xanthoxyloides (G. Don) DC.

1. F. sogdiana Bge. in Mem. Ac. Sci. St.-Petersb. Sav. Entrang. 7 (1854) 390; Boiss. Fl. or 4 (1879) 41; V. Vasil'ev in Fl. SSSR, 18 (1952) 502; Fl. Kazakhst. 7 (1964) 92; Opred. rast. Sr. Azii [Key to Plants of Mid. Asia] 8 (1986) 31; Fl. Sin. 61 (1992) 39.

—Ic.: Fl. SSSR, 18, Plate 25, fig. 10; Fl. Kazakhst. 7, Plate 10, Fig. 5; Fl. Sin. 61, tab. 11, fig. 1-2.

37  Described from East. Kazakhstan. Type in Moscow (MW).

Along valleys and river banks, forms floodplain forests and groves. Quite often, in plantations in oases.

IIA. Junggar: *Tien Shan* (valley of Ili river 2 km east of Yamatu crossing, along foot of main bank on left, Aug. 21, 1957—Yun., Lee et Yuan').

IB. Kashgar: *Nor.* (Bai oasis, outside settlement, Sept. 21, 1957—Yun., Lee et Yuan'). General distribution: Fore Balkh., Jung.-Tarb., Nor. Tien Shan; Mid. Asia.

2. F. xanthoxyloides (G. Don) DC. Prodr. 8 (1844) 275; Clarke in Hook. f. Fl. Brit. Ind. 3 (1882) 606; Fl. W. Pakistan, 59 (1974) 3; Fl. Xizang. 3 (1986) 874; Fl. Sin. 61 (1992) 33.— *Ornus xanthoxyloides* D. Don, Gen. Hist. Dichlam. Pl. 4 (1837) 57.

—Ic.: Fl. W. Pakistan, 59, fig. 1 c–d; Fl. Sin. 61, tab. 9, fig. 5.

Described from West. Himalayas. Type in London (BM).

Along river banks, floors of gorges, rocky slopes, about 3000 m.

IIIB. Tibet: *South.* (Ali; "Chzhada" — Fl. Xizang, l.c.).

General distribution: Fore Asia, India, Nor. Afr.

## 2. Syringa L.
## Sp. pl. (1753) 9

1. Leaves simple, entire ....................................................................... 2.
+ Leaves pinnatisected ............................. 2. S. pinnatifolia Hemsl.
2. Leaves glabrous, broad-cordate, with faintly emarginated or truncated base. Capsule glabrous ...................... 1. S. oblata Lind.
+ Leaves pubescent, oval, with cuneate base. Capsule verrucose ...
.................................................................... 3. S. pubescens Turcz.

1. S. oblata Lindl. in Gard. Chorn. 1859 (1859) 869; Franch. Pl. David. 1 (1884) 205; Forbes et Hemsl. Ind. Fl. Sin. 2 (1889) 83; Walker in Contribs U.S. Nat. Herb. 28 (1941) 650; Pl. vasc. Helanshan (1986) 199; Drev. rast. Tsinkhaya [Woody Plants of Qinghai] (1987) 539; Fl. Sin. 61 (1992) 71; Fl. Intramong. 4 (1993) 73. — *S. giraldii* Lemoine, Catal. No. 155 (1903) 8. — *S. giraldiana* Schneider ex Diels in Filchner Wissensch. Ergebn. 10, 2 (1908) 262.

— Ic.: Fl. Sin. 61, tab. 20, fig. 1–2; Fl. Intramong. 4, tab. 27, fig. 3–4.

Described from East. China (Shanghai). Type in Cambridge (CGE)?

On exposed slopes among shrubs and in undergrowth, along flanks and floors of gorges, river shoals, up to 2400 m. Extensively cultivated as an ornamental plant.

IA. Mongolia: *Alash. Gobi* (Alashan mountain range; midportion of west. slope, June 20–July 10, 1873 — Przew.; vicinity of Ninsya-fu town, April 28, 1909 — Napalkov; vicinity of Baisy monastery, 2100 m, juniper thickets, July 6, 1957 — Petr.; "Wang-yeh-fu, No. 38; Shui-mo-kou, No. 97; Hala-hou-kou, No. 51–May, 1923, Ching" — Walker, l.c.; "on slopes of Helanshan mountain range" — Fl. Intramong. l.c.).

IIIA. Qinghai: *Nanshan* (Kazhi village on Karyn river [Gumbum district, Lachisan' mountain range], May 5, 1885 — Pot.; "Datunkhe" — Drev. rast. Tsinkhaya [Woody Plants of Qinghai] l.c.), *Amdo* (mountains west of Dzhamba river, March 14, 1885 — Pot.).

General distribution: China (Dunbei, Nor., Nor-West., East., South-West. north).

Note. Var. *alba* Hort with white flowers and var. *giraldii* (Lemoine) Rehder are found independently and together with type.

2. S. pinnatifolia Hemsl. in Gard. Chron., ser. 3, 39 (1906) 68; Pl. vasc. Helanshan (1986) 197; Drev. rast. Tsinkhaya [Woody Plants of Qinghai] (1987) 537; Fl. Sin. 61 (1992) 79. — *S. pinnatifolia* var. *alashanensis* Ma et S.Q. Zhou in Fl. Intramong. 5 (1980) 412 and ed. 2, 4 (1993) 76.

— Ic.: Pl. vasc. Hellanshan, tab. 36; Fl. Sin. 61, tab. 21, fig. 4–5; Fl. Intramong. 4, tab. 28, fig. 5.

Described from South-West. China (Sichuan). Type in London (K).

In undergrowth and among shrubs on mountain slopes, 2600–3100 m.

IA. Mongolia: *Alash. Gobi* (Alashan mountain range: "Helanshan" — Fl. Intramong. l.c.; Pl. vasc. Helanshan, l.c.).

IIIA. Qinghai: *Amdo* (found in eastern border region — Fl. Sin. l.c.; Drev. rast. Tsinkhaya [Woody Plants of Qinghai] l.c.).

General distribution: China (Nor.-West.-Gansu, South-West-.-Sichuan west.). Cultivated as an ornamental plant.

3. S. pubescens Turcz. in Bull. Soc. Natur. Moscou, 13 (1840) 73; Drev. rast. Tsinkhaya [Woody Plants of Qinghai] (1987) 542; Fl. Sin. 61 (1992) 63. — *S. villosa* auct. non Vahl (1804); Franch. Pl. David. 1 (1884) 204; Hook. f. in Curtis' Bot. Mag. 115 (1889) tab. 706; Forbes et Hemsl. Index Fl. Sin. 2 (1902) 83.

— Ic.: Fl. Sin. 61, tab. 18, fig. 1–7.

Described from Nor. China (vicinity of Beijing). Type in St.-Petersburg (LE).

On mountain slopes, gorges and river banks among shrubs, up to 3000 m.

IIIA. Qinghai: *Nanshan* ("in lower course of Datun-khe" — Drev. rast. Tsinkhaya [Woody Plants of Qinghai] l.c.), *Amdo* (oppidum Guidui, 2100 m, May 7, 1885 — Pot.).

General distribution: China (Dunbei, Nor., Nor.-West., Cent., South-West.: Sichuan nor.).

Note. The small-leaved variety regarded by some investigators as a subspecies or independent species is widely distributed within the distribution range of the species: *S. pubescens* Turcz. var. *tibetica* Batalin in Acta Horti Petrop. 13 (1894) 118; Schneid. in Feddes Repert. 9 (1910) 80. — *S. pubescens* Turcz. subsp. *microphylla* (Diels) M.C. Chang et X.L. Chen in Invest. Stud. Nat. 10 (1990) 34; M.Chang in Fl. Sin. 61 (1992) 66. — *S. microphylla* Diels in Bot. Jahrb. 29 (1900) 531; Schneid. Ill. Handb. Laubh. 2 (1911) 778. — *S. potaninii* Schneid. in Feddes Repert. 9 (1910) 80; Stapf in Curtis, Bot. Mag. 150 (1924) tab. 9060. This species is also often cultivated as an ornamental plant.

## 3. Jasminum L.
### Sp. pl. (1753) 7

1. J. officinale L. l.c.; Dc. Prodr. 8 (1844) 313; Clarke in Hook. f. Fl. Brit. Ind. 3 (1882) 603; Rehd. in Sargent. Pl. Wils. 2 (1916) 613; Hand.-Mazz. Symb. Sin. 7 (1936) 1013; Hara in Enum. flow. pl. Nepal, 3 (1982) 81; Fl. Xizang, 3 (1986) 890; Fl. Sin. 61 (1992) 192.

— Ic.: Curtis, Bot. Mag. 1, tab. 31; Fl. Xizang, 3, tab. 342, fig. 4–9; Fl. Sin. 61, tab. 51, fig. 1–3.

Described from India (Himalayas). Type in London (Linn.).

On mountain slopes, talus, among shrubs, 3800–4000 m.

IIIB. Tibet: *South.* ("nor of Lhasa" — Fl. Xizang, l.c.).

General distribution: China (South-West.), Himalayas and all over Mediterr. up to south. and Atlant. Europe.

Note. The dwarfish, small-leaved variety *J. officinale* L. var. *tibeticum* C.Y. Wu ex P.Y. Bai in Acta Bot. Yunnan, 1 (1979) 155, fig. 5; Fl. Xizang, 3 (1986) 890, tab. 342, fig. 9, grows in Tibet proper and has been described from around Lhasa.

## Family 95. BUDDLEJACEAE Wilh.

### 1. Buddleja L.

Sp. pl. (1753) 112

1.   Leaves alternate .......................................................................... 2.

+   Leaves opposite ......................................................................... 4.

2.   Leaves 3–10 cm long, lanceolate, acute, entire or gently crenate. Capsule glabrous .................................... 1. B. alternifolia Maxim.

+   Leaves very small, 3–30 and up to 50 mm long, oval, short-cuspidate. Capsule pubescent with stellate hairs ...................... 3.

3.   Leaves very small, 3–6 mm long, entire, pubescent on both sides with stellate hairs, compact .............................. 2. B. minima Bao.

+   Leaves 3–5 cm long, entire or large-toothed in upper half, glabrous, spaced ................................................. 3. B. wardii Marq.

4.   Leaves ovate or oblong-ovate with cordate base, sinuate-dentate and crispate along margin. Flowers in uppermost compact racemose inflorescence with small poorly visible bracts ................................................................... 4. B. crispa Benth.

+   Leaves hastate with gently sinuate or truncated base, rough—toothed. Flowers in sessile axillary glomerules with bracts larger than latter ................................. 5. B. hastata Prain ex Marq.

1. B. alternifolia Maxim. in Bull. Ac. Sci. St.-Petersb. 26 (1880) 494; Forbes et Hemsl. Index Fl. Sin. 2 (1889) 119; Walker in Contribs U.S. Nat. Herb. 28 (1941) 651; Fl. Xizang, 3 (1986) 895; Pl. vasc. Helansh. (1986) 20; Drcv. rast. Tsinkhaya [Woody Plants of Qinghai] (1987) 552; Fl. Sin. 61 (1992) 269; Fl. Intramong. 4 (1993) 77.

—Ic.: Curtis' Bot. Mag. 151, tab. 9085; Fl. Xizang, 3, tab. 343; fig. 5; Fl. Sin. 61, tab. 71, fig. 1–9; Fl. Intramong. 4, tab. 28, fig. 1–4.

Described from Nor.-West. China (Gansu). Type in St.-Petersburg (LE).

On sunny, mountain slopes, in ravines and along banks of rivers, among shrubs, up to 2500 m.

IA. Mongolia: *Alash. Gobi* (Alashan mountain range: Hsi-Jeh-kou, Holanshan Mountains, alt. 1875 to 2100 m, No. 185, May 10–25, 1923—Ching), *Ordos* ("Otokachi" —Fl. Intramong, l.c.).

IIIA. Qinghai: *Nanshan* (On Itel'-gol river, mid-April 1885—Pot.; "Min'khe" —Dev. rast. Tsinkhaya [Woody Plants of Qinghai] l.c.).

General distribution: China (Nor.—Hebei, Nor.-West., Cent.—Henan, South-West.).

2. B. crispa Benth. Scroph. Ind. (1835) 43; DC. Prodr. 10 (1846) 444; Forbes et Hemsl. Index Fl. Sin. 3; Hand.-Mazz. Symb. Sin. 7 (1936) 947; Hara et al. Enum. flow. pl. Nepal, 3 (1982) 89; Fl. Xizang. 3 (1986) 898;

Fl. Sin. 61 (1992) 298.— *B. tibetica* W.W. Smith. in Rec. Bot. Surv. Ind. 4 (1911) 270.

—Ic.: Curtis' Bot. Mag. 80, tab. 4793; Fl. Xizang, 3, tab. 344, fig. 1–4; Fl. Sin. 61, tab. 81, fig. 1–4.

40     Described from Nor. India. Type in London (K).

Along arid slopes of mountains, 3800–4300 m.

IIIB. Tibet: *South.* ("Lhasa, Shigatsze Naidun"—Fl. Xizang, l.c.).

General distribution: China (Nor.-West.—Gansu, South-West.), Himalayas, Fore Asia.

3. B. hastata Prain ex Marq. in Kew Bull. 1930, 197; Fl. Xizang, 3 (1986) 898; Fl. Sin. 61 (1992) 295.

—Ic.: Fl. Xizang, 3, tab. 345, fig. 1–3; Fl. Sin. 61, tab. 79, fig. 4–7.

Described from East. Himalayas. Type in London (K).

Along slopes of mountains, among shrubs, up to 3800 m.

IIIB. Tibet: *South.* ("Lhasa"—Fl. Xizang, l.c.).

General distribution: East. Himalayas.

4. B. minima S.Y. Bao in Fl. Xizang, 3 (1986) 895, tab. 343, fig. 6–8.

Described from Tibet (Lhasa). Type in Kunming (KUN).

Along arid mountain slopes, 3100–3800 m.

IIIB. Tibet: *South.* ("Lhasa, Shigatsze", "Lhasa, No. 75321—C.Y. Wu et al—typus !" —Fl. Xizang, l.c.).

General distribution: China (South-West.—Sikang).

Note. This species was included without any explanation in Flora of People's Republic of China (Fl. Sin. 61, 271) as a synonym of *B. alternifolia* Maxim. I have not examined specimens of this species but, judging from the brief diagnosis and its sketch, it differs very greatly from *B. alternifolia* Maxim. (pubescent capsule, very small compact leaves) for being placed among synonyms of the latter species.

5. B. wardii Marq. in J. Linn. Soc. (London) Bot. 48 (1929) 203 and in Kew Bull. (1930) 185; Fl. Xizang, 3 (1986) 895; Fl. Sin. 61 (1992) 271.— *B. tsetangensis* Marq. l.c. 202.

—Ic.: Fl. Sin. 61, tab. 71, fig. 10.

Described from South. Tibet. Type in London (K).

Along rocks and on river shoals, 3000–3400 m.

IIIB. Tibet: *South.* ("Naidun"—Fl. Xizang, l.c.).

General distribution: endemic.

# Family 96. GENTIANACEAE Juss.

1.     Flowers 4-merous; corolla with 4 spurs ........... 8. Halenia Borkh.

+     Flowers 5- and 4-merous; corolla without spurs ........................ 2.

2.     Corolla with distinctly manifest tube, not shorter than its lobes ....................................................................................... 3.

+ Corolla divided almost down to base into lobes and its tube many times shorter than latter, poorly visible ............................ 5.

3. Corolla claviform with long thin tube and 5-lobed limb, bright pink or white; ovary and capsule subbilocular; post-anthesis anthers coiled .................................................. 1. Centaurium Hill.

+ Corolla tubular-campanulate, tubular-infundibular, tubular or campanulate, of different colours; ovary and capsule unilocular; post-anthesis anthers not coiled .................................................. 4.

4. Plants with twining or creeping stem and axillary pendent campanulate 4-merous blue flowers ...........................................
.................................. 2. Pterygocalyx Maxim. (*P. volubilis* Maxim.).

+ Plant with erect or ascending stems but never twining or creeping. Flowers upright, 5- or rarely 4-merous, mostly in terminal inflorescences or terminal flowers single ........................
.................................................................... 3. Gentiana L.

5. Plants successively dichotomously branching from base; flowers pale green or whitish, nutant, small, about 5 mm long, 1–3 each in leaf axils ................................................................
.................. 6. Anagallidium Griseb. (*A. dichotomum* (L.) Griseb.).

+ Plants with simple or branched stems with opposite branches. Flowers usually very large and bright-coloured, erect, very rarely nutant ................................................................. 6.

6. Style not developed and stigma sessile, in the form of 2 bands running downward along ovarian sutures; nectaries poorly visible; flowers blue, single, on long pedicels. Annuals ............ 7.

+ Style distinctly; manifest and stigma capitate or bilobate; nectaries distinctly manifest, mostly fimbriate along margin; flowers usually in fascicles of 2 or 3 unequal pedicels in inflorescences. Perennials and annuals .................... 7. Swertia L.

7. Lobes of corolla inside at base with nectar cavity, without appendage ................................................. 4. Lomatogonium A. Br.

+ Lobes of corolla inside at base with large scarious entire blade ............................ 5. Lomatogoniopsis T.N. Ho et S.W. Liu.


## 1. Centaurium Hill
### Brit. Herb. (1756) 62

1. Flowers on stalks of varying length in lax, forked inflorescence, rarely flowers single ..................... 1. C. pulchellum (Sw.) Druce.

+ Flowers subsessile on one side of lateral branches, forming loose spicate inflorescences ................ 2. C. spicatum (L.) Fritsch.

1. C. pulchellum (Sw.) Druce, Fl. Oxf. (1897) 342; Grossh. in Fl. SSSR, 18 (1952) 528; Fl. Kirgiz. 8 (1959) 185; Fl. Kazakhst. 7 (1964) 35; Grub. Opred. rast. Mong. [Key to Plants of Mongolia] (1982) 200; Opred. rast. Sr. Azii [Key to Plants of Mid. Asia] 8 (1986) 33; Fl. Sin. 62 (1988) 10.— *Gentiana pulchella* Sw. in Vet. Acad. Handl. 4 (1783) 85. tab. 3, fig. 8–9. — *Centaurium meyeri* (Bge.) Druce in Rep. Bot. Exch. Cl. Brit. Isles, 1916 (1917) 613; Kryl. Fl. Zap. Sib. 9 (1937) 2168; cum auct. Grossh.; Grossh. in Fl. SSSR, 18 (1952) 529; Fl. Kirgiz. 8 (1959) 186; Fl. Kazakhst. 7 (1964) 96; Fl. Intramong. 5 (1980) 69; Pl. vasc. Helanshan (1986) 200.— *Erythraea meyeri* Bge. in Ldb. Fl. alt. 1 (1829) 220; Persson in Bot. Notis. (1938) 297. — *E. pulchella* Fries, Novit. Fl. Suec. ed. 2 (1828) 74; Danguy in Bull. Mus. nat. hist. natur. 20 (1914) 75.— *E. pulchella* β *albiflora* Ldb. Fl. Ross. 3 (1846) 51.— *E. ramosissima* Pers. var. *altaica* Griseb. in DC. Prodr. 9 (1845) 57; Forbes et Hemsl. Index Fl. Sin. 2 (1890) 126.— *Centaurium pulchellum* var. *altaicum* (Griseb.) Kitag. et Hara in J. Jap. Bot. 13 (1937) 26; Kitag. Lin. Fl. Mansh. (1939) 357; Fl. Sin. 62 (1988) 12.— *Gentiana centaurium* L. Sp. pl. (1753) 230.

—Ic.: Ldb. Ic. pl. Fl. Ross. 2, tab. 159; Reichb. Ic. Fl. Germ. 17, tab. 1061; Fl. Sin. 62, tab. 2, fig. 1–5.

42    Described from Scandinavia. Type in Stockholm (S).

In wet and solonetzic coastal meadows, sandy shoals, among bulrushes in tugais, along borders and banks of irrigation ditches in ploughed lands.

IA. Mongolia: *East. Mong.* (right bank of Huang He below Hekou town, Aug. 4; Huang He valley, Aug. 9; Termin-Bashin area, Aug. 10–1884, Pot.), *Depr. Lakes* (Chovd-gol Aue an den Ongochyuul, Aug. 17, 1976—Hilbig, Schamsran; nor.-east. bank of Khara-Us-Nur lake, valley of Chono-Kharaikh stream connecting Khara-Nur and Khara-Us-Nur lakes, Sept. 3, 1978—Gub.), *Val. Lakes* ("Tuin-gol Aue bei Bogd, 1979"—Hibig et Schamsran, l.c.), *Alash. Gobi* (Khara-Sukhai area on Edzin river, July 28, 1886—Pot.), *Ordos* (Dzhasygen-Qaidam area, Aug. 30, 1884—Pot.; 25 km south-east of Otokachi town, Khablaitunao lake, Aug. 1; 10 km west of Ushinchi town, on bank of pool, Aug. 4; 20 km west of Dzhasakachi town, in valley, Aug. 16–1957, Petr.).

IB. Kashgar: *Nor.* (Khaidyk-gol river, Chubogorin-nor area, 1200 m, Aug. 17, 1893—Rob.; Uchturfan, Yaman-Su, June 9, 1908—Divn.; between Maralbashi and Aksu at Chadyr-kul' settlement, Aug. 6, 1929—Pop.; Temyntsan' in Karashar, 1200 m, No. 6886, June 29; Maralbashi area nor-west of Shasai-khuduk, No. 7786, Sept. 2–1958, Lee et Chu), *West.* (Bakh village on Charlym river, Aug. 6, 1909—Divn.; Kashgar oasis, Upal village, along river bed, July 12, 1929—Pop.; "Bostan-terek, about 2400 m, Aug. 14, 1934"—Persson, l.c.), *East.* (south. fringe of Khami oasis, Bugas village, 480 m, Aug. 22, 1895—Rob.; in Toksun region 7 km nor.-east of Ilakhu lake, 150 m alt., No. 7297, June 16; 4 km nor. of Yuili, No. 8527, Aug. 2; Yuili area, Dantazhen'fantszy village, No. 8587, Aug. 9; Yuili area, in Chiglik alongside drying up lake, No. 8595, Aug. 12–1958, Lee et Chu).

IIA. Junggar: *Cis-alt.* ("Altai, montagnes entre Oulioungur et Kobdo, Sept. 5, 1895 — Chaff." — Danguy, l.c.), *Tien Shan* (on Tekes river, 3350 m, July 10, 1893 — Rob.; "entre le Sairam-Nor et l'Ebi-Nor, July 27, 1895 — Chaff." — Danguy, l.c.), *Jung. Gobi* (Nom area, Aug. 28, 1895 — Rob.; 3-4 km east of St. Kuitun settlement on Shikho-Manas road, Saza zone, July 7, 1957 — Yun., Lee, Yuan'; "Dzakhoi-Dzaram oasis" — Grub. l.c.), *Zaisan* (Kaba river in the vicinity of Kaba village, June 16; between Kara area and Kaba village, on bank of irrigation ditch, June 16–1914, Schischk.).

IIIA. Qinghai: *Nanshan* (Lovachen town on bank of Sinin-he river branch, July 26, 1908 — Czet.).

General distribution: Aralo-Casp., Fore Balkh., Jung.-Tarb., Nor. and Cent. Tien Shan; Europe, Balk,–Asia Minor, Caucasus, Mid. Asia, West. Sib. (Altay), China (Altay, Dunbei, Nor.).

2. **C. spicatum (L.) Fritsch** in Mitt. Naturwiss. Ver. Univ. Wien, 5 (1907) 97; Grossh. in Fl. SSSR, 18 (1952) 535; Fl. Kirgiz. 8 (1959) 186; Fl. Kazakhst. 7 (1964) 97; Opred. rast. Sr. Azii [Key to Plants of Mid. Asia] 8 (1986) 34. — *Gentiana spicata* L. Sp. pl. (1753) 230. — *Erythraea spicata* Pers. Syn. 1 (1805) 283; Ldb. Fl. Ross. 3 (1846) 51.

— Ic.: Reichb. Ic. Fl. Germ. 17, tab. 1061; Fl. SSSR, 18, Plate 27, fig. 1; Fl. Kazakhst. 7, Plate 11, fig. 3.

Described from South. Europe. Type in London (Linn.).

In moist solonetzic coastal meadows, tugais.

IB. Kashgar: *East.* (southern fringe of Khami oasis, Bugas village, 480 m, among bulrushes, No. 294, Aug. 19, 1895 — Rob.).

General distribution: Aralo-Casp., Fore Balkh.; South. Europe, Mediterr., Balk.-Asia Minor, Fore Asia, Caucasus, Mid. Asia.

Note. As in *C. pulchellum,* in this species too, white-flowered form is found rather often along with pink-flowered form; moreover, these 2 forms often grow together.

## 2. Pterygocalyx Maxim
### Prim. Fl. Amur. (1858) 198

1. **P. volubilis Maxim.** Prim. Fl. Amur. (1858) 198 and 274; Fl. Intramong. 5 (1980) 67; Pl. vasc. Helanshan (1986) 200; Fl. Sin. 62 (1988) 311. — *Crawfurdia volubilis* (Maxim.) Makino in Bot. Mag. Tokyo, 4 (1890) 86; Kitag. Lin. Fl. Mansh. (1939) 358; Grossh. in Fl. SSSR, 18 (1952) 537.

— Ic.: Maxim. l.c. tab. 9; Fl. Intramong. 5, tab. 27, fig. 5–8.

Described from Far East (Fore Amur). Type in St.-Petersburg (LE).

In scrubs, among undergrowth and forest borders.

IA. Mongolia: *Alash. Gobi* (Alashan mountain range — Pl. vasc. Helanshan, l.c.).

IIIA. Qinghai: *Nanshan* (Sin'chen town surroundings, 2150–2450 m, coiling on shrubs, Aug. 28, 1901 — Lad.).

General distribution: Far East (south. and Sakhalin), China (Dunbei, Nor., Nor.-West., Cent., South-West.), Korean peninsula, Japan.

### 3. Gentiana L.

Sp. pl. (1753) 227; Froelich, Gentiana (1790) 19; Bge. in Nuov. Mem. Soc. Natur. Moscou, 1 (1829); Griseb. Gentian. (1839); Kusn. in Trav. Soc. natur. St.-Petersb. 24 (1894) 3

1. Lobes of corolla separated from each other by intermediate folds-lobes of smaller size and different form; calyx with distinct tube. Perennials and annuals (subgenus 1. *Gentiana*) ... 2.

+ Folds-lobes absent between lobes of corolla and adjoin each other directly; calyx deeply laciniated into lobes, with very short tube. Annuals or rarely biennials with thin root (subgenus 2. Gentianella (Moench) Kusn.) ..................................................... 69.

2. Perennials with developed rhizome and funiform roots or with contracted rhizome and filamentous root ................................. 3.

+ Annuals, mostly small plants, with thin rachiform root ......... 40.

3. Calyx with broad winglike keels along nerves; corolla tubular, with broad rotate limb, intense blue, 32–47 mm long; stigma entire, funnel-shaped. Flowers single on long stem with 1–2 pairs of small leaves. Leaves in compact radical rosette, broad-ovoid (section 11. Cyclostigma Griseb.) ....................................................
.......................................................................... 79. G. uniflora Georgi.

+ Calyx without winglike keel; corolla funnel-shaped or campanulate, with erect or slightly declinate lobes; stigma bipartite with linear lobes. Leaves on radical rosette generally narrow, linear to ovoid-lanceolate ............................................. 4.

4. Lower cauline leaves form radical rosette or in rosettes on vegetative shoots ........................................................................... 8.

+ Radical rosette of leaves absent; lower cauline leaves small or undeveloped scalelike; upper large. Flowers aggregated in small numbers and surrounded by leaves at tip of simple stem ........................................................................................... 5.

5. Leaves with 3 or 5–7 nerves. Flowers aggregated at tip of stem and surrounded by reduced leaves. Folds of corolla asymmetrical, fimbriate (section 3. Pneumonantha (Gled.) Gaudin) ..................................................................................... 6.

+ Leaves with a midnerve. Flowers usually in 2's, surrounded by large leaves. Folds of corolla symmetrical, entire or dentate (section 4. Isomeria Kusn.) ............................................................. 7.

6. Leaves lanceolate, attenuated into long cusp, with 3 nerves. Corolla dark blue, its lobes obtuse ...... 31. G. dschungarica Rgl.

+ Leaves ovoid or oblong-ovoid, obtuse, with 5–7 nerves. Corolla blackish blue, its lobes acute ................. 32. G. fischeri P. Smirn.

7.   Stem erect, simple, 7–15 cm tall; leaves ovoid-lanceolate, sessile on broad and high funnel-shaped sheaths. Lobes of calyx rather oblong, erect; upper part of corolla blue, lower part yellowish with blue bands .................................... 33. G. amplicrater Burk.

+    Stem procumbent, 4–6 cm long, branched; leaves and lobes of calyx obovoid, with chondroid margin and declinate tip. Corolla lilac-red or bluish lilac ................ 34. G. urnula H. Smith.

8.   Small plants (5–10, rarely up to 20 cm tall), forming small turfs; stems simple, many, ascending or procumbent, sometimes contracted and rather few; leaves linear or linear-lanceolate. Flowers large, single, terminal ........................................ 9.

+    Plants not forming turfs; stems developed, single or more, mostly simple, branched in inflorescence, erect or ascending. Flowers in terminal inflorescences, rarely single ..................... 17.

9.   Folds of corolla symmetrical, large. Leaves linear, chondroid along margins and with short spiny cusp. Corolla dark violet-blue, with 5 green bands outwardly, narrow-campanulate, (30) 40–45 (55) mm long. Seeds glabrous (section 6. Chondrophylla Bge.) ...................................... 38. G. grandiflora Laxm.

+    Folds of corolla asymmetrical, short. Leaves lanceolate; if linear, soft, without chondroid margin and spiny cusp. Seeds with alveolate surface (section 2. Frigida Kusn., subsection Monopodiae (H. Smith) T.N. Ho ............................................. 10.

10.  Leaves opposite .......................................................... 11.

+    Leaves in whorls 6 each; stems many, ascending; rhizome contracted with fibrous root .................................................
     .............................................. 23. G. hexaphylla Maxim. ex Kusn.

11.  Cauline leaves 1–2 pairs, small; leaves of radical rosette quite large, 3–6 cm long, lanceolate, chondroid along margins; rhizome long and thick, with spaced roots; stems rather few, short .................................................................................. 12.

+    Cauline leaves many, compact, linear or lanceolate, 7–15 (20) mm long; stems many; root fibrous ............................................. 13.

12.  Corolla blue, without spots but with white bands along nerves; teeth of calyx broad-lanceolate ............. 25. G. szechenyi Kanitz.

+    Corolla yellowish white or blue with many dark spots; teeth of calyx narrow-lanceolate ........... 19. G. callistantha Diels et Gilg.

13.  Cauline leaves narrow-linear, slender ....................................... 14.

+    Cauline leaves lanceolate or broad-lanceolate ......................... 15.

14.  Corolla dark blue, blue with dark spots at throat .....................
     ........................................................ 22. G. futtereri Diels et Gilg.

+    Corolla light blue, yellowish, without spots at throat ...................
...................................................................... 21. G. farreri Balf. f.

    15.     Cauline leaves and teeth of calyx white-chondroid along margins and with cusp; rhizome developed. Corolla greyish light blue, rarely white with broad grey-blue bands and sometimes with spots ............................... 24. G. stipitata Edgew.

+     Cauline leaves and teeth of calyx without chondroid emargination and cusp; rhizome not developed ...................... 16.

16.     Flowers 6–7 cm long; calyx 3/5–2/3 of corolla length, its teeth longer than tube; corolla light blue ...............................................
.................................................................. 20. G. dolichocalyx T.N. Ho.

+     Flowers 3–6 cm long; calyx only 1/3–1/2 of corolla length, its teeth shorter than tube; corolla dark blue ......................................
................................................................ 26. G. veitchiorum Hemsl.

17.     Stems enveloped at base with fibrous remnants of sheaths of radical leaves. Corolla with limb or its lobes erect; folds of corolla symmetrical (section 1. Cruciatae Gaudin) .................. 21.

+     Stems at base without fibrous cover. Corolla without limb, with erect lobes, its folds asymmetrical (section 2. Frigida Kusn., subsection Sympodia H. Smith) .................................................. 18.

18.     Flowers in inflorescence 2–6 each, rarely single; corolla light yellow or golden white with dark speckles or bands. Stems usually 17–25 cm tall .................................................................... 19.

+     Flowers 1–2 each at tip of stem. Corolla dark blue or pale yellow, without speckles and bands. Stem usually 8–15 cm tall ............................................................................................... 20.

19.     Corolla pale yellow with bluish grey transverse bands ...........
................................................................ 30. G. purdomii Marq.

+     Corolla golden white with violet specks ......... 27. G. algida Pall.

20.     Corolla dark blue, somewhat intumescent under throat. Stem usually 8–15 cm tall; leaves and teeth of calyx scabrous along margins .................................... 29. G. przewalskii Maxim.

+     Corolla pale yellow, infundibular-tubular, not intumescent. Stem contracted. mostly 6–8 (10) cm tall; leaves and teeth of calyx glabrous along margins ................. 28. G. nubigena Edgew.

21.     Inflorescence lax, racemose, sometimes flowers compact only at tip, rarely single; all flowers mostly on distinctly manifest pedicels ..................................................................................... 22.

+     Inflorescence compact, corymbose-capitate; flowers many, sessile ......................................................................................... 33.

32.  Stems procumbent, ascending at tip. Flowers 1–3 (5) each at tip of stem; single flowers axillary on long (up to 5 cm long in case of axillary flowers) pedicels; corolla infundibular-tubular, 28–38 mm long; calyx with 5 unequal, 1–3 (5) mm long teeth, scarious with green nerves ..................................... 5. G. gracilipes Turrill.

+  Stems erect or steeply ascending. Inflorescence terminal, many-flowered; corolla tubular-infundibular, about 30 mm long; calyx green with 5 more or less equal, 5–7 mm long teeth ......................
.................................................. 15. G. tianschanica Rupr.

33.  Radical leaves poorly developed, small; cauline leaves broad, enlarging toward tip of stem, ovoid-lanceolate, sessile. Inflorescence covered by leafy ochrea. Stems strong, thick .... 34.

+  Radical leaves large, cauline leaves much smaller, reducing toward tip of stem. Inflorescence not covered by leafy ochrea ........................................................................................ 35.

34.  Tube of corolla 26–28, rarely 22, mm long, broad, yellow-green inside, with violet-brown longitudinal bands outside; limb yellow ........................................... 16. G. tibetica King ex Hook. f.

+  Tube of corolla 20–22 mm long, narrow, yellow, with dark spots at throat, limb blue or lilac ......... G. crassicaulis Duthie ex Burk.

35.  Corolla pale yellow or yellow-green; calyx less than half of corolla ....................................................................................................... 36.

+  Corolla dark blue or blue-violet; calyx less than half of corolla, with very short teeth ..................................................................... 39.

36.  Calyx entire, tubular, its teeth distinctly manifest, 3–8 mm long; corolla tubular; ovary pedicellate. Radical leaves linear to lanceolate ....................................................................................... 37.

+  Calyx lanciniated on one side, with very short teeth (up to 1 mm long) or without distinctly manifest teeth; ovary sessile. Radical leaves broad, ovoid-lanceolate or broad-lanceolate..38.

37.  Teeth of calyx linear-lanceolate, 6-8 mm long; corolla 25–30 mm long, 5–7 mm in diam. at throat ........................................
.................................................. 18. G. walujewii Rgl. et Schmalh.

+  Teeth of calyx subulate or linear, 3–4 or 5–7 mm long; corolla 30–35 mm long, 7–10 (12) mm in diam. at throat ..........................
............................................................. 10. G. olgae Rgl. et Schmalh.

38.  Calyx deeply laciniated, its teeth compact and poorly visible; corolla without bands and spots, campanulate, 21–38 mm long ................................................. 12. G. robusta King ex Hook. f.

+  Calyx short-laciniated, truncated, with 5 very short spaced teeth; corolla with blue longitudinal bands and spots, tubular,

18–20 (25) mm long. Lower cauline leaves with highly connate sheaths ................................................... 9. G. officinalis H. Smith.

39.    Corolla 18–20 mm long; ovary sessile. Radical leaves broad, oblong-lanceolate or lanceolate .............. 8. G. macrophylla Pall.

+    Corolla 23–26 mm long; ovary pedicellate. Radical leaves linear or broad-linear ................... 13. G. siphonantha Maxim. ex Kusn.

40.    Leaves and teeth of calyx without chondroid border; radical leaves very small and do not form rosette; cauline leaves, however, increase gradually toward tip of stem; if, however, leaves are same-sized, plant with fairly tall erect stem. Folds of corolla asymmetrical; style very long, equal to ovary. Capsule without terminal crest, oblong-oval .......................................... 41.

+    Leaves and teeth of calyx mostly with chondroid border; radical leaves largest and usually gathered in a rosette; cauline leaves, however, decrease gradually toward tip. Corolla mostly with limb, its folds symmetrical; style considerably shorter than ovary. Capsule long-pedicellate and quite often with crest at tip. Small annuals (section 6. Chondrophylla Bge.) ................ 43.

41.    Stem erect, 10–30 cm tall, tetragonal, branched in upper half, many-flowered. Corolla tubular-infundibular without limb, 4–6 cm long, milky white with black longitudinal bands; calyx with subulate teeth (section 5. Stenogyne Franch.) ............................... ................................................................. 37. G. striata Maxim.

+    Small plants 3–8 cm tall, branching from base with procumbent shoots. Corolla with limb, clavate, up to 15 mm long; calyx with truncated tube and implanted broad-ovoid teeth, sharply tapering into pedicel (section 2. Frigida Kusn., subsection Annua (Marq.) Grub.) .................................................. 42.

42.    Corolla pale yellow with blue speckles ......................................... ................................................................. 35. G. tongolensis Franch.

+    Corolla light blue, without speckles ........... 36. G. vernayi Marq.

43.    Plants branched right from base, their shoots simple, terminating in single flower; cauline leaves with cusp or obtuse but not rounded tip. Teeth of calyx invariably erect ................ 44.

+    Plant with developed stem, branching in upper half, or with simple one- or many-flowered; if branched from base, shoots branching again, each with one or more flowers. Teeth of calyx erect or declinate ........................................................................ 65.

44.    Capsule oblong or linear-terete, its length a few time more than breadth ................................................................................... 45.

| | |
|---|---|
| + | Capsule globose, obovoid or oval; if at all its length exceeds breadth, it is by not more than 1.5–2 times ............................... 52. |
| 45. | Leaves in upper part of shoots (above centre) larger than in lower part. Calyx funnel-shaped, its teeth flat, scarious ......... 46. |
| 48 + | Leaves in upper part of shoots comparatively smaller than in lower part. Calyx tubular, its teeth keeled, rigid; pedicels short ....................................................................................... 48. |
| 46. | Corolla white, with light blue bands and spots outside, 11–13 mm long ......................................... 43. G. caeruleo-grisea T.N. Ho. |
| + | Corolla blue or lilac ...................................................................... 47. |
| 47. | Shoots strong, 6–12 cm tall. Corolla 18–25 mm long, funnel-shaped, somewhat lilac-coloured above, yellow-green at base, without bands; capsule coriaceous, not exserted from corolla ................................................................................ 57. G. pudica Maxim. |
| + | Shoots weak, slender, 1.5–3 cm tall. Corolla 12–15 mm long, tubular, light blue. Capsule scarious, exserted from corolla ........ ................................................................... 48. G. hyalina T.N. Ho. |
| 48. | Leaves imbricate in 4 rows, broad-ovoid with cusp. Corolla lilac-coloured, with green longitudinal bands outside .................. .............................................................. 64. G. tetrasticha Marq. |
| + | Leaves in pairs, spaced, obovoid ................................................. 49. |
| 49. | Calyx half of tubular light blue corolla; stamens with very short filaments at throat of corolla .................... 51. G. ludlowii Marq. |
| + | Calyx up to 3/4 of corolla length .............................................. 50. |
| 50. | Plant 1–2 cm tall; shoots with 2–3 pairs of leaves. Seeds winged ..................................... 46. G. crenulato-truncata (Marq.) T.N. Ho. |
| + | Plants larger, 3–6 (10) cm tall; leaves many. Seeds without wings ........................................................................................ 51. |
| 51. | Leaves on margins, especially at base, scabrous, with papillae. Capsule tapers gradually conelike into pedicel ........................... ................................................................... 49. G. karelinii Griseb. |
| + | Leaves glabrous along margins. Capsule sharply tapered at base, orbicular. Flower usually arcuate .......... 54. G. nutans Bge. |
| 52. | Cauline leaves linear or narrow-lanceolate ............................... 53. |
| + | Cauline leaves spatulate to oval and ovoid .............................. 57. |
| 53. | Leaves and teeth of calyx coriaceous with chondroid margins. Calyx tubular with short teeth; corolla narrow-tubular, 7–9 mm long, 1.5–2 mm in diam. at throat, dark blue ............................... .............................................................. 53. G. micantiformis Burk. |

+ Leaves and teeth of calyx herbaceous with scarious margins ............................................................................................ 54.

54. All leaves, cauline as well as radical, narrow-linear or linear-lanceolate with long awn at tip ................. 41. G. aristata Maxim.

+ Radical leaves elliptical or broad-ovoid; cauline leaves linear-lanceolate or lanceolate, but all leaves with short cusp ........... 55.

55. Corolla lilac-coloured, with broad dark lilac bands outside; similar speckles at throat. Radical leaves elliptical .....................
................................................................ 62. G. syringea T.N. Ho.

+ Corolla dark blue, grey, white or variegated. Radical leaves broad-ovoid .............................................................................. 56.

56. Corolla grey or white, with yellow tube and dark speckles at throat. Cauline leaves compactly sessile, imbricate .....................
................................................................ 47. G. grumii Kusn.

+ Corolla dark blue in upper part, lower part yellow-green with broad green bands, without specks at throat ...............................
................................................................ 65. G. tricolor Diels et Gilg.

57. Leaves and teeth of calyx with chondroid clear margins and short rigid, often declinate cusp ................................................. 58.

+ Leaves and teeth of calyx without chondroid margin or rigid cusp ..................................................................................... 62.

58. Leaves and teeth of calyx ciliolate along margins; stems fine-scabrous. Corolla white with greenish longitudinal bands ..........
................................................................ 55. G. prattii Kusn.

+ Leaves and teeth of calyx glabrous and smooth along margins. Corolla light blue or lilac ................................................. 59.

59. Flowers sessile; corolla 7–9 mm long, light blue, its folds bidentate. Stems often simple, one-flowered ...............................
................................................................ 44. G. clarkei Kusn.

+ Flowers distinctly pedicellate; corolla 9–14 mm long ............. 60.

60. Calyx with 5 broad white keels, tubular, only 1/3 shorter than corolla; folds of corolla entire .............. 42. G. burkillii H. Smith.

+ Calyx with poorly discernible keels, tubular-infundibular, half of corolla ................................................................................ 61.

61. Corolla 9–12 mm long, light blue, concealed, its lobes acute; folds orbicular, obtuse, denticulate. Capsule highly exserted from corolla. Leaves obovoid or spatulate with orbicular tip ......
................................................................ 56. G. pseudo-aquatica Kusn.

+ Corolla 12–14 mm long, light blue or pink, broadly exposed, its lobes obtuse, folds acute, deltoid, entire or short-bidentate.

Capsule not exserted from corolla. Leaves spatulate with deltoid tip ............................................... 60. G. spathulifolia Kusn.

62. Cauline leaves and teeth of calyx with broad scarious margins. Stem poorly branched at base. Corolla light blue, its lobes acuminate, with callosity at tip ......... 63. G. tatsienensis Franch.

+ Cauline leaves and teeth of calyx without broad scarious border. Lobes of corolla without callosity at tip ....................... 63.

63. Corolla white, with dark lead-coloured longitudinal bands, with dark speckles at throat; its folds fringe-toothed ....................
............................................... 50. G. leucomelaena Maxim.

+ Corolla light blue; its folds bipartite, dentate or almost entire .....
................................................................................................. 64.

64. Corolla concealed, 9–10 mm long; its folds almost entire or denticulate. Capsule highly exserted from corolla .......................
............................................................... 40. G. aquatica L.

+ Corolla broadly exposed, 9–12 mm long, its folds bipartite. Capsule not exserted from corolla .............. 39. G. aperta Maxim.

65. Stem mostly single, erect, branched in upper half; leaves spatulate, with orbicular tip without cusp, imbricate. Teeth of calyx invariably erect, deltoid; notches between them acute .......
.................................................................. 58. G. riparia Kar. et Kir.

+ Plants branched from base. Teeth of calyx mostly declinate, orbicular or obovoid, with short rigid cusp; notches between them invariably obtuse, orbicular ............................................... 66.

66. Leaves sessile, with thickened chondroid margins .................. 67.

+ Leaves on distinctly manifest petioles, longer than orbicular blade, with thin margin, Corolla light blue, at throat with bands, yellow-green tube; its lobes acute, serrulate .......................
........................................................ 52. G. mailingensis T.N. Ho.

50  67. Stems with fine glandular hairs; leaves of radical rosette well developed, orbicular, 6–30 mm long; cauline leaves spatulate, like teeth of calyx, with cusp recurved outward. Calyx intumescent; corolla dark blue ................... 61. G. squarrosa Ldb.

+ Stem without glandular pubescence; radical leaves not developed. Calyx long-tubular, not intumescent, with declinate teeth; corolla light blue ................................................ 68.

68. Leaves in lower part of stem orbicular; in upper part, like teeth of calyx, reniform, glabrous. Corolla twice longer than calyx ......
............................................... 45. G. crassuloides Bureau et Franch.

+ Cauline leaves and teeth of calyx elliptical or ovoid, acute, with divaricate hairs along margins at base. Corolla only slightly longer than calyx ...................................... 59. G. simulatrix Marq.

69.  Corolla with a ring of fimbrilla at throat ................................. 70.

\+  Corolla without fimbrilla inside tube ........................................ 75.

70.  Corolla funnel-shaped with limb and compact ring of fimbrilla, 5-, rarely 4-lobed, ovary and capsule sessile. Biennial plants, 20–40 cm tall, with (4) 6–10 pairs of leaves (section 9. Endotricha Froel.) .................................................. 73. G. acuta Michx.

\+  Corolla tubular or tubular-infundibular, mostly without limb, with erect lobes; each of them at base with 1 or 2 fimbriate glumes, 4- or 5-lobed (section 8. Comastoma Wettst.) ........... 71.

71.  Stem developed, erect, branched in upper half, 8–20 cm tall ...................................................................................... 72.

\+  Plants branched from root neck, with many simple ascending one-flowered stems, 4–8 (15) cm tall ........................................ 73.

72.  Stem 4-sided. Lobes of calyx with crispate margins: corolla lilac-red, its lobes ovoid, obtuse, declinate .............................................. ...................................................... 70. G. polyclada Diels et Gilg.

\+  Stem orbicular. Lobes of calyx flat; corolla blue-violet with pale tube, its lobes subdeltoid, acute, erect ............................................. .................................................................... 71. G. pulmonaria Turcz.

73.  Flowers large, 12–18 mm long, 5-merous; corolla dark-violet — blue, with highly declinate lobes ................. 68. G. falcata Turcz.

\+  Flowers small, 6–10 (but up to 18 during fruiting) mm long; 5- or 4-merous; corolla of different colour, with erect lobes ...... 74.

74.  Corolla light blue with thin orange-coloured nerves; its lobes orbicular, obtuse, without callosity at tip ........................................ ........................................................................ 72. G. tenella Rottb.

\+  Corolla lilac-red, its lobes ovoid with callosity at tip ................... ......................................................... 69. G. pedunculata Royle ex G. Don.

75.  Lobes of corolla along margins, specially at base, fimbriate, rarely almost altogether glabrous. Flowers large, 25–40 (65) mm long, 4-merous (section 7. Crossopetalum Froel.) ................... 76.

\+  Corolla altogether glabrous, without fimbrilla and cilia. Flowers mostly small, 4–13 mm, rarely up to 19 mm long, 4–5-merous (section 10. Arctophila Griseb.) ................................................. 77.

76.  Stem erect, simple or poorly branched in upper half; cauline leaves linear or linear-lanceolate. Two teeth of calyx as long as tube of corolla or even longer, linear-subulate, and almost twice longer than 2 other acute-deltoid teeth; corolla bright blue with oblong-ovoid lobes with rather few fimbrillae in lower part along margins ................................................. 66. G. barbata Froel.

+ Stem branched from base, rarely simple, erect. Calyx barely reaches middle of corolla tube, its teeth identical, broad-deltoid, acute; corolla light blue or lilac, its lobes broad-obovoid, with many short fimbrillae. Leaves ovoid-lanceolate or oblong-oval ................................................ 67. G. paludosa Munro ex Hook. f.

77. Annual-biennial plants 10–30 cm tall; stem erect, branched in upper part, with whorls of flowers in axils of upper leaves. Corolla 9–13 (19) mm long, bluish violet, light yellow or white, its lobes with cusp; lobes of calyx linear-lanceolate, somewhat uneven and hamate at tip; notches between them obtuse ............ ............................................ 78. G. turkestanorum Gand.

+ Small annuals, 2–8 cm tall; stem branched shrublike from base. Flowers single, terminal; corolla 4–7, rarely up to 13 mm long, its lobes obtuse ........................................................... 78.

78. Leaves on distinctly manifest petioles, spatulate or obovoid. Flowers 4-merous, 4–5 mm long ................................................. 79.

+ Leaves sessile. Flowers 5-merous, (5) 7–13 mm long; corolla sky-blue ............................................................................ 80.

79. Corolla lilac-coloured; lobes of calyx spatulate, declinate at tip ................................................ 74. G. arenaria Maxim.

+ Corolla light yellow; lobes of calyx oblong-oval, erect ................. ............................................ 77. G. pygmaea Rgl. et Schmalh.

80. Stems glabrous. Leaves and lobes of calyx ovoid-lanceolate, notches between lobes of calyx acute; lobes of calyx with black margins. Plant 3–10 cm tall ............................... 75. G. azurea Bge.

+ Stems with glandular hairs. Leaves and lobes of clayx crispate along margins, notches between them obtuse. Plant about 3 cm tall ...................................... 76. G. moorcroftiana Wall. ex Griseb.

## Subgenus 1. Gentiana
## Section 1. Cruciatae Gaudin

1. G. biflora Rgl. ex Kusn. in Acta Horti Petrop. 13, 2 (1893) 62 and 15, 3 (1904) 321; Marq. in Kew Bull. 3 (1937) 167.—*G. dahurica* auct. non Fisch.: Fl. Sin. 62 (1988) 64, p.p., quoad syn.

Described from plant raised in St.-Petersburg Botanical Garden from seeds collected by N. Przewalsky in Nanshan. Type in St.-Petersburg (LE).

In high-mountain meadows?

IIIA. Qinghai: *Nanshan* (Kansu occidentali, Przewalski semina legine dicitus 1884 Januario [in northern Momo-Shan' mountain range]. Ex Horto bot. Petropol. July 1888 —typus !).

General distribution: endemic ?

Note. In all characteristics, undoubtedly belongs to section *Cruciatae* Gaudin and is closely affiliated to *G. dahurica* Fisch., but differs sharply in paired 4-merous flowers, pedicellate ovary and contracted stem with rosette of broad radical leaves. Possibly mutant?

52      **G. crassicaulis** Duthie ex Burk. in J. Asiat. Soc. Bengal, new ser. 2 (1906) 311; H. Smith in Hand.-Mazz. Symb. Sin. 7 (1936) 978; Marq. in Kew Bull. 3 (1937) 168; Fl. Xizang, 3 (1986) 919; Fl. Sin. 62 (1988) 67. — *G. tibetica* auct. non King; Forbes et Hemsl. Index Fl. Sin. 2 (1902) 136.

—Ic.: Fl. Xizang, 3, tab. 349; Fl. Sin. 62, tab. 10, fig. 1–4.

Described from South-West. China (Yunnan). Type in London (K).

In alpine meadows and meadowy slopes, among shrubs, in forest meadows and borders, along sides of mountain roads, 2100-4500 m.

IIIA. Qinghai: Amdo (?—Fl. Sin. l.c.).

IIIB. Tibet: *South.*?

General distribution: China (Nor.-West.: south. Gansu, South-West.).

Note. We have no reliable information about the find of this species within Central Asia but the possibility is not ruled out since *G. crassicaulis* is distributed in the proximate border regions of southern Gansu and Kam.

2. **G. dahurica** Fisch. in Mem. Soc. Natur. Moscou, 3 (1812) 63; Palibin in Acta Horti Petrop. 14 (1895) 131; Kusn. ibid, 15 (1898) 318, p.p. excl. syn.; Marq. in Kew Bull. 3 (1937) 166; Grossh. in Fl. SSSR, 18 (1952) 566; Grub. Konsp. fl. MNR [Conspectus of Flora of Mongolian People's Republic] (1955) 223; Fl. Intramong. 5 (1980) 72; Grub. Opred. rast. Mong. [Key to Plants of Mongolia] (1982) 201; Pl. vasc. Helanshan (1986) 201; Fl. Sin. 62 (1988) 64, p.p. excl. syn.

—Ic.: Fl. Intramong. 5, tab. 29; Grub. Opred. rast. Mong. [Key to Plants of Mongolia] Plate 108, fig. 486.

Described from East. Siberia (Dauria). Type in St.-Petersburg (LE).

In cereal grass and forbs-cereal grass steppes, on rocky steppe slopes.

IA. Mongolia: *Cent. Khalkha* (cent. Kerulen near Bars-Khoto, mid-Aug.; same site, near San-beis duke's site, Aug.—1899, Pal.; 15 km west of Tsagan-Obo somon centre, Aug. 5, 1956—Dashnyam), *East. Mong.* (between Ulan-Khada and Qindai, Aug. 10; Kalgan gorge, Aug. 23–1898, Zab.; Ulan-Dzhilgu area, July 3; between Lykse lake and Ulugui river, July 5–1899, Pot. and Sold.?; Manchuria railway station, Aug. 22, 1902— Litw.; 55 miles south-east of Durkhe-arshan, Aug. 13, 1927—Terekhovko; Khalkhin-gol river, Symbur area, Sept. 1, 1928—Tug.; Zodol-Khan-Ula, May 14, 1944—Yun.; 50–55 km east of Erentsab, Shavorte-Nur area, Aug. 19, 1949—Yun.; vicinity of Khabarga railway station north of Enger-Shanda, Aug. 1; 30 km north of Choibalsan somon centre, Aug. 12–1954; 5 km north of Enger-Shanda, south-east. slope of mountain, Aug. 15; Delger-Munkh mountain north of Choibalsan somon centre, mountain foothills, Aug. 29–1959, Dashnyam; vicinity of Shilin-Khoto, steppe, 1959—Ivan.; Dariganga, Shiliin-Bogdo-ula, nor. slope, Aug. 10; 11 km south of Khamar-Daba settlement on road to Numurgin-gol, left bank of Khalkhin-gol, Aug. 15–1970, Grub., Ulzij. et al; 37 km nor.-west of Khutliin-khuduk, huommocky area, July 14, 1971—Dashnyam, Karam., Safronova; 13 km south—

south-east of Khamar-Daba settlement, July 26, 1971 — Karam. et Safronova), *East. Gobi* (east of Shara-Murun, alt. 5000 ft, No. 466, 1925 — Chaney), *Alash. Gobi* (Alashan mountain range, Yamata gorge, south. slope, upper belt, May 7, 1908 — Czet.?).

General distribution: East. Sib. (south. Dauria), Nor. Mong. (Mong.-Daur., Cis-Hing.), China (Dunbei, Nor.).

Note. The affinity of Potanin and Soldatov's specimens from East. Mongolia to this species arouses some doubt since calyx in them has long teeth but laciniated on one side. The treatment of sterile plants collected by Czetyrkin in Alashan mountain range as *G. dahurica* is rather unreliable.

53     3. **G. decumbens** L. f. Suppl. (1781) 174; Griseb. Gen. et sp. Gentian. (1839) 279 and in DC. Prodr. 9 (1845) 110; Ldb. Fl. Ross. 3 (1847) 64; Franch. Pl. David. 1 (1884) 211; Kusn. in Acta Horti Petrop. 15 (1898) 295; Danguy in Bull. Mus. nat. hist. natur. 20 (1914) 75; Kryl. Fl. Zap. Sib. 9 (1937) 2186; Kitag. Lin. Fl. Mansh. (1939) 358; Grossh. in Fl. SSSR, 18 (1952) 562; Grub. Konsp. fl. MNR [Conspectus of Flora of Mongolian People's Republic] (1955) 223; Fl. Kazakhst. 7 (1964) 102; Grub. Opred. rast. Mong. [Key to Plants of Mongolia] (1982) 201; Pakhom. in Opred. rast. Sr. Azii [Key to Plants of Mid. Asia] 8 (1986) 40; Fl. Sin. 62 (1988) 61. — *G. adscendens* Pall. Fl. Ross. 2 (1788) 106, tab. 54. — *G. gebleri* Ldb. ex Bge. in Mem. Soc. Natur. Moscou, nov. ser. 1 (1829) 211, tab. 8.

— Ic.: Fl. Kazakhst. 7, Plate 11, fig. 8.

Described from Siberia. Type in London (Linn.).

On steppe and meadowy slopes, in steppe and solonetzic coastal meadows, in petrophyte steppes, rocky slopes and talus, on coastal pebble beds, 1000–2500 m.

IA. Mongolia: *Khobd.* (Altyn-cheche, July 9, 1870 — Kalning; bank of Tsagan-gol river 30 versts–1 verst = 1.067 km — from its inflow into Kobdo river, July 29; Ukha river, Aug. 6–1899, Lad.; Kharkhira mountain group, steppe west of Mal. Khara-Usu-nor, July 18; same site, south of Namyur river, July 18; same site, Nitsegun river, tributary of Namyur river, July 21; same site, Mostek area south-west of Ulangom, July 25–1903, Gr.-Grzh.; Achit-nur lake basin in interfluvine area of Bukhu-Muren and Khub-Usu-gol, 3 km east of Bukhu-Muren somon centre, sandy pebble bed of floodplain, July 15, 1971 — Grub., Dariima et al; floodplain of Bukhu-Muren river at same-named somon, July 24, 1977 — Karam., Sanczir et al), *Mong. Alt.* (Taishiri-ula, July 15, 1877 — Pot.; "steppe des environs de Kobdo, Sept. 25, 1895, Chaff." — Danguy, l.c.; between Khulmu lake and Nam-daban pass, Aug. 4; in Ogyul'dzo daban, Aug. 5, 1896; left bank of Senkir river, July 17; in Mal. Ulan-daba pass, July 18; bank of Zusulun brook, left tributary of Bulugun, July 25–1898, Klem.; Dain-Gol lake, central hillock, July 12, 1908 — Sap.; Khasagtu-Khairkhan mountains, bank of Dundu-Tseren-gol river, Sept. 17, 1930 — Pob.; Taishiri-ula mountain range, nor. slope 15 km south-east of ajmaq-administrative territorial unit in Mongolia — centre, lower belt, along floors of creek valleys, July 11, 1945; lower Turgen-gol river, left bank tributary of Bulgan river, July 7; nor. trail of Khan-Taishiri mountain range, July 14; 2–3 km south of Tamchi lake, in valley, July 17; nor. trail of Bus-Khairkhan mountain range, ravine, July 17; valley of Indertiin-gol at summer camp of Bulgan somon, July 24; 25–30 km south of Tamchi-daba pass, midcourse of Bidzhiin-gol, valley slope, under birch grove near spring, Aug. 10–1947, Yun.; Bidzhiin-gol gorge 7–8 km from estuary, in creek valley along brook, Sept. 8, 1948 — Grub.; Khasagtu-Khairkhan mountain range, Tsagan-Irmyk-ula, nor. slope in upper Khunkerin-ama, 2500 m, Aug. 23, 1972 — Grub.,

Ulzij. et al; basin of Tonkhil-nur lake, lower Dzuilin-gol, Aug. 10; valley of Arshantyn-gol [Uenchin-gol basin] 3 km from estuary, 2400 m, Aug. 14–1979, Grub., Dariima, Muld.), *Cent. Khalkha* (cent. Kerulen near camp of Dalai-beis duke, 1899–Pal.; Ubur-Dzhargalante river, between sources and Agit mountain, Aug. 10; same site, second terrace, Aug. 11; same site, rubbly slopes, Aug. 31; near Orgochen-sume monastery, Aug. 23; subalpine zone of Ubur-Dzhargalante left bank between Bogota and Agit mountains, Sept. 15; between sources of Ubur-Dzhargalante river and Achit mountain, Sept. 22–1925, Krasch. et Zam.; vicinity of Ikhe-Tukhum-nor lake, Mishik-gun-estuary of Temeni-ama, June 1926; vicinity of Nalaikha mine, Aug. 2, 1927–Zam.), *East. Mong.* (Borogol, May 31, 1898–Zab.; Dariganga, Shiliin-Bogdo-ula, nor. slope in upper portion, Aug. 10; 10 km west of Shiliin-Bogdo-ula near foothill of basalt conical hillock, Aug. 10–1970, Grub., Ulzij. et al), *Depr. Lakes* (vicinity of Kholbo-nur lake, Khudzhirte south-east of right bank of Burgasutai river, July 27; south. bank of Khara-Usu lake, Aug. 18; Torkhula mountains on Kobdo river, Aug. 24; Ulangom, Sept. 5–1879, Pot.; 10 km nor.–nor.-west of Ulangom on road to Khara-Usu, highest liman–drowned river valley–Aug. 13; 4–5 km west of Ulangom, Aug. 17–1931, Bar.; Khobdossk state farm in lower Buyantu river, 1941 and 1944–Kondratenko; valley of Kobdo river at crossing of Kobdo-Ulangom highway, Aug. 23, 1944–Yun.), *Val. Lakes* (right bank of Ongiin-gol facing urton, July 28, 1893; between Olon-nor and Taisheri-Ula mountains, July 17, 1894–Klem.; not far from Dzak-Baidarik, Aug. 6, 1930–Pob.; right bank of Tatsin-gol river 8 km north of Narin-Tel' somon, July 30, 1952–Davazhamts), *Gobi Alt.* (Dundu-Saikhan mountains, west. conical hillock, midbelt, July 7, 1909–Czet.; same site, on steppe slope, Aug. 19, 1931–Ik.-Gal.; Dundu- and Dzun-Saikhan mountain ranges, July-Aug. 1933–M. Simukova; Dundu-Saikhan mountain range, south. slope under main peak, 2824 m, along crest, July 23, 1970–Grub., Ulzij. et al).

IIA. Junggar: *Cis-Alt.* ("entre l'Ouchte et l'Irtich, alt. 1900 m, Aug. 20; entre Oulioungour et Kobdo, Sept. 9–1895, Chaff."–Danguy, l.c.; left bank of Kran river under Urmogaity pass, Sept. 22, 1876–Pot.; Qinhe dist., in forest, No. 1225, Aug. 2; Qinhe dist. between Kun'tai and Chzhunkhaitsza, 2500 m, No. 1105, Aug. 5; Chzhun-khaitsza, 2400 m, in forest, No. 1471, Aug. 7; Koktogai dist., lower course of Khasyungou river, 1800 m, No. 2062, Aug. 18–1956, Ching), *Tien Shan* (Tor-Kul' area [Turkul'], end of Aug. 1895–Kozlov), *Jung. Gobi* (Khoni-Usuni-Gobi, Gun-Tamga area, Aug. 2, 1947–Yun.; valley of Bulgan-gol 7 km below Ulyastain-gol estuary along road at Bulgan somon centre, floodplain, in chee grass thicket, Aug. 28, 1979–Grub., Dariima, Muld.).

General distribution: Fore Balkh., Jung-Tarb., Nor. Tien shan (Trans-Ili Ala Tau); Europe (Volga-Kam., Trans-Volga), West. Sib. (Altay, upper Irtysh), East. Sib., China (Altay?, Nor.: Hebei).

4. G. fetissowii Rgl. et Winkl. in Acta Horti Petrop. 7, 2 (1881) 548; Kusn. in Acta Horti Petrop. 15 (1904) 324; Grossh. in Fl. SSSR, 18 (1952) 567; Fl. Kazakhst. 7 (1964) 103; Pakhom. in Opred. rast. Sr. Azii [Key to Plants of Mid. Asia] 8 (1986) 40. – *G. macrophylla* var. *fetissowii* Ma et K.C. Hsia in Acta Sci. natur. univers. Intramong. 6, 1(1964) 43; Fl. Sin. 62 (1988) 74, p.p. quoad pl. songar. – *G. macrophylla* auct. non Pall.: Ldb. Fl. Ross. 3, 1 (1847) 69, p.p. quoad pl. ex Ala Tau.

–Ic.: Gartenfl. 31, tab. 1069.

Described from plants raised in St.-Petersburg Botanical Garden from seeds collected by A. Fetisow in Mal. Yuldus in East. Tien Shan. Type in St.-Petersburg (LE).

On steppe and meadowy slopes, forest meadows and glades, in alpine and subalpine meadows, Cobresia thickets, 1800–3500 m.

IIA. Junggar: *Jung. Alat.* (in Ven'tsyuan' dist., 2640 m, No. 4623, Aug. 5; mountains in Toli region, No. 2629, Aug. 7–1957, Kuan; same site, 3 km from Balaktau, subalpine meadow, No. 1050, Aug. 6, 1957–Shen; Dzhair mountain range, Dzhair pass on road from Shikho to Chuguchak, mountain steppe belt, Aug. 9, 1957–Yun., S.-y. Lee, Yuan'), *Tien Shan* (Urtak-sary, July 20, 1878–Fet.; upper Borotal, 1850–2150 m, No. 1878; upper Algoi, 2450 m, Sept. 11. 1879–A. Reg.; B. Yuldus, 2450–2750 m, alpine meadow, Aug. 6, 1893–Rob.; nor. slopes of Tien Shan, Daban 2300 m, No. 1969, July 18, 1957–Kuan; 7–8 km south of Danu, 3200 m, No. 526, July 22, 1957–Shen; Manas river basin, upper Danu-gol near estuary of lateral creek valley leading to Danu pass, sedge-*Cobresia* meadow, July 21, 1957–Yun., S.-y. Lee, Yuan').

General distribution: Fore-Balkh. (mountainous), Jung.-Tarb.

Note. The species is very close to *G. macrophylla* Pall. and yet quite individualistic.

5. **G. gracilipes** Turrill in Curtis' Bot. Mag. 141 (1915) pl. 8630; Marq. in J. Roy. Hort. Soc. 57 (1932) 204; id. in Kew Bull. 3 (1937) 166.—*G. kurroo* var. *brevidens* Maxim. ex Kusn. in Bull. Ac. Sci. St.-Petersb. 34 (1892) 508; Diels in Futterer, Durch Asien (1903) 13.—*G. dahurica* auct. non Fisch.: Kusn. in Acta Horti Petrop. 15 (1898) 398, p.p. incl. syn.; Rehder in J. Arn. Arb. 14 (1933) 28; Marq. in Kew Bull. 3 (1937) 166; Walker in Contribs U.S. Nat. Herb. 28 (1941) 651; Pl. vasc. Helanshan (1986) 201 ?; Fl. Sin. 62 (1988) 64, p.p.—*G. tianschanica* β *kozlowii* Kusn. in Acta Horti Petrop. 15 (1898) 304.—*G. decumbens* auct. non L.f.: Hao in Bot. Jahrb. 68 (1938) 629.—*G. pseudodecumbens* H. Smith ex Marq. in Kew Bull. 3 (1937) 130.

—Ic.: Curtis' Bot. Mag. 141, pl. 8630.

Described from plants raised in Kew Botanical Garden from seeds collected in Gansu province (Nor.-West. China)-Type in London (K).

On steppe and meadowy slopes of mountains, coastal meadows and shoals, spruce and juniper forests, among shrubs, 1000–4000 m.

IA. Mongolia: *East. Mong.* (Muni-ula, exposed mountain summits, Aug. 18, 1871–Przew.; valley of Doristai river and Shukhan-gol, on mountain ranges and gorges between them, Aug. 20; valley of Ulan-Morin river, Aug. 22; Chagan-nor lake, Aug. 26 –1884, Pot.), *Alash. Gobi* (Alashan mountains, Sept. 8, 1871; same site, July 14, 1873–Przew.; "Holanshan, on grasslands, No. 1090, 1923, B.C. Ching"–Walker, l.c.), *Ordos* (18 km from Dzhasakacha town at Chingiz-khan mausoleum, flat hillock, Aug. 15, 1957 –Petr.), *Khesi* (Loukhu-Shan' mountain range, south-east. and south-west. slopes, July 17, 1908–Czet.).

IC. Qaidam: *Mount.* (Dulan-khit temple, 3350 m, in exposed glades in spruce and juniper forests, Aug. 9; same site, in estuary of Karagaityn-ama gorge, Aug. 9–1901, Lad.).

IIIA. Qinghai: *Nanshan* (northern marginal mountain range, steppe [July 2] 1872; North. Tetung mountain range, alpine belt, Aug. 2, 1872; on Kuku-nor lake, 3100 m, on shoals, July 14; South. Tetung mountain range, Aug. 6–1880, Przew.; on Itel'-gol river, Aug. 1885–Pot.; South. Tetung mountain range, July 28, 1908–Czet.; Bingou village, Aug. 25, 1908–Czet.; 25 km south of Gulan town, east. extremity of Nanshan, Aug. 12, 1958; Kuku-nor lake, 3200 m, meadow on east. bank, Aug. 5; 33 km west of Xining town,

2450 m, rocky slopes of hillocks, Aug. 5; 66 km west of Xining, 2800 m, rocky slopes of hillocks, Aug. 5; 20 km west of Gunhe, basin, 2980 m, Aug. 6–1959, Petr.; "oberhalb Tankar-thing, Futterer" — Diels, l.c.; "Koko-nor, Kiapokia, 3200–3400 m, 1930" — Hao, l.c.; "Tai Hua, on exposed, dry slopes and along clay roadsides, No. 529, 1923, R.C. Ching" — Walker, l.c.), *Amdo* ("grasslands between Labrang and Yellow River, Rock" — Render, l.c.; "oberhalb Balekun-gomi, Futterer" — Diels, l.c.).

IIIB. Tibet: *Weitzan* (Burkhan-Budda mountain rang, nor. slope Khatu gorge and Ikhe-gol, 3650–3950 m, on clayey descents of mountains, July 22, 1901 — Lad.; "south. bank of Russkoe lake, on mountains and hillocks, 4100 m, July 29-31, 1884 — Przew." — Kusn. l.c., typus *G. tianschanica* β *kozlowii* Kusn.).

General distribution: China (Nor., Nor.-West., South-West. — nor.).

Notes 1. The species is very similar to *G. tianschanica* Rupr. The colour of corolla varies from dark blue to dark violet, occasionally with white spots. Calyx sometimes entire, sometimes laciniated rather shallowly on one side and its teeth short, unequal and of variable length.

2. Neither type nor any other specimens of *G. tianschanica* β *kozlowii* Kusn. could be found in the Herbarium of the Komarov Botanical Institute.

6. G. kaufmanniana Rgl. et Schmalh. in Acta Horti Petrop. 7, 1 (1880) 331; Kusn. in Acta Horti Petrop. 5 (1898) 317; Persson in Bot. notis. (1938) 297; Grossh. in Fl. SSSR, 18 (1952) 565; Fl. Kirgiz. 8 (1959) 190; Fl. Kazakhst. 7 (1964) 103; Pakhom. in Opred. rast. Sr. Azii [Key to Plants of Mid. Asia] 8 (1986) 41; Fl. Sin. 62 (1988) 66.

— Ic.: Fl. Kazakhst. 7, Plate 12, fig. 1; Fl. Sin. 62, tab. 9, fig. 9–11.

Described from Nor. Tien Shan. Type in St.-Petersburg (LE).

In alpine meadows, sparse spruce forests and among dwarf juniper, old moraines, syrts (upland watersheds) and sasas (solonchaks on lateral seepages) of glaciers, on rocks and talus, 1800–3500 m.

IB. Kashgar: *West.* ("Jerzil-jajlek, 3500 m, July 12, 1931" — Persson, l.c,).

IIA. Junggar: *Jung. Alt.* (Aksu dist., Ven'tsyuan' town, on slope, No. 3476, Aug. 14, 1957 — Kuan), *Tien Shan* (Urten-Muzarta gorge, Aug. 4, 1877 — Fet.; Aktash in Dzhagastai mountains, 2450–2750 m, Aug. 11, 1877 — A. Reg.; Yuldus, Sept. 1878 — Fet.; Aryslan river [tributary of Kash river], 2450–2750 m, July 8; same site, July 17; Kunges river, Aug. 27; Dzagastai-gol [head of Yuldus] 2750–2900 m, Sept. 6–1879, A. Reg.; Kapsalan mountains, Kara-dzhyun area, up to 2450 m, July 17; near Kapsalan pass, 2750–3050 m, July 18–1893, Rob.; Passe zwischen Kin-su and Kurdai, July 3, 1907 — Merzb.; Uch-Turfan, Uitasy from Karashar toward Bedel' pass, June 27, 1908 — Divn.; "Atchelek pass, about 2900 m, Aug. 9, 1932" — Persson, l.c.; 20 km south of Ven'tsyuan' town, 2810 m, No. 1503, Aug. 14; same site, 2900–3200 m, No. 1537, Aug. 14–1957, Shen; 10 km north of Chzhaosu, 2980 m, No. 3301, Aug. 15, 1957 — Kuan; 3 km south of Yakou town [Barchat], 2900 m, No. 1691, Aug. 30; north of Yakou town, No. 1732, July 3–1957, Shen; Nilki dist., 60 km north of Ulyastai, 2850 m, No. 3999, Aug. 31, 1957 — Kuan; Hetzin dist., Mal. Yuldus, 2900 m, alluvial valley, No. 6310, Aug. 2; Narat mountains along Kunges, 2300 m, No. 6556, Aug. 7; Tsagan-nur pass in Khotun-Sumbul, 3100 m, in valley alluvium, No. 6511, Aug. 11–1958, Lee and Chu; Narat mountain range, south. slope along road from Dasht pass to Bain-Bulak, along ravines, Aug. 8; basin of Mal. Yuldus 8 km short of Kotyl' pass on road to Karashar, Aug. 15–1958, Yun. et Yuan'; on Kucha-Yakka-Aryk road, 2850 m, No. 10008, July 26; Kucha town, 2590 m, intermontane basin, No. 10060, July 26–1959, Lee and Chu).

General distribution: Jung.-Tarb., Nor. and Cent. Tien Shan; Mid. Asia (Pam.-Alay).

56

7. G. lhassica Burk. in J. Asiat. Soc. Bengal, nov. ser. 2 (1906) 311; Marq. in Kew Bull. 3 (1937) 184; Fl. Xizang, 3 (1986) 915; Fl. Sin. 62 (1988) 63.

—Ic.: Fl. Xizang, 3, tab. 348, fig. 10–12.

Described from South. Tibet. Type in London (K).

In alpine meadows and meadowy slopes, coastal meadows, 3500–4900 m.

IIIB. Tibet: *Weitzan* (Yantszy-tszyan basin, vicinity of Chzherku monastery and along Tsza-chyu river, 3600 m, along mountain slopes, No. 475 a, b and c, Aug. 16; same site, vicinity of Chzherku monastery, 3600 m, No. 476, Aug. 16–1900, Lad.; "Bizhu" — Fl. Xizang, l.c.), *South.* ("In valle rivuli Kyi-chu dicti, prope Lhasa, No. 1642, Walton" — typus ! — Burk. l.c.).

General distribution: China (South-west.: Kam).

8. G. macrophylla Pall. Fl. Ross. 2 (1788) 108; Ldb. Fl. Ross. 3, 1 (1847) 69, 70, excl. pl. Alatav.; Kryl. Fl. Zap. Sib. 9 (1937) 2185; Kitag. Lin. fl. Mansh. (1938) 359; Grossh. in Fl. SSSR, 18 (1952) 567; Grub. Knsp. fl. MNR [Conspectus of Flora of Mongolian People's Republic] (1955) 224; Fl. Kazakhst. 7 (1964) 104; Fl. Intramong. 5 (1980) 72; Grub. Opred. rast. Mong. [Key to Plants of Mongolia] (1982) 73; Fl. Sin. 62 (1988) 73, p.p., excl. var. *fetissowii.* — *G. jacutensis* Bge. in DC. Prodr. 9 (1845) 118.

—Ic.: Pall. l.c. tab. 46; Fl. SSSR, 18, Plate 29, fig. 3; Fl. Kazakhst. 7, Plate 12, fig. 2; Fl. Intramong. 5, tab. 30; Fl. Sin. 62, tab. 10, fig. 5–7.

Described from Siberia. Type in London (BM) or in Berlin (B)?

In forbs steppes, dry valley, forest, floodplain, marshy and alpine sedge-*Cobresia* meadows, larch forests and their borders, 800–2400 m.

IA. Mongolia: *Khobd.* (between Dzusylan and Sarincha, south. slope in forest belt, July 15, 1879 — Pot.; Turgen mountain range, 7 km above Turgen-gol estuary, terrace above floodplain, sedge-*Cobresia* meadow, July 17, 1971 — Grub., Dariima, Ulzij.; valley of Turgen river, 2300 m, sedge-*Cobresia* meadow with shrubs, July 9, 1973 — Banzragch, Karam. et al), *Mong. Alt.* ("Environs de Kobdo, steppe, Aug. 22, 1895), Chaff." — Danguy, l.c.), *Mid. Khalkha* (Ubur-Dzhargalante river head, marshy meadow, Aug. 10, 1925 — Krasch. et Zam.), *East. Mong.* (Shilin-Khoto, true steppe, 1959 — Ivan.; Mongolia chinensis in reditu e China, 1841 — Kirilow).

General distribution: West. Sib. (south., Altay), East. Sib., Far East (south.), Nor. Mong., China (Dunbei, Nor., Nor.-West.).

9. G. officinalis H. Smith in Hand.-Mazz. Symb. Sin. 7 (1936) 979; Marq. in Kew Bull. 3 (1937) 167; Walker in Contribs U.S. Nat. Herb. 28 (1941) 652; Fl. Sin. 62 (1988) 72. — *G. macrophylla* var. *albolutea* Limpr. in Feddes Repert., Beih. 12 (1922) 466. — *G. macrophylla* auct. non Pall.: Marq. l.c. 167, p.p.; Pl. vasc. Helanshan (1986) 201? — *G. fetissowii* auct. non Rgl. et Winkl.: Marq. l.c. 167.

Described from South-West. China (Sichuan). Type in Wien (Vienna).

In wet and coastal meadows, 2300–4200 m.

IA. Mongolia: *Alash. Gobi* (Alashan mountain range: midportion of west. slope, alpine meadow, July 17, 1873 — Przew.?; Khote-gol gorge, nor. slope, midbelt, June 19, 1908 — Czet.?; "Holanshan, No. 1063, 1923, R.C. Ching" — Walker, l.c.).

IIIA. Qinghai: *Amdo* ("Nima-Lang-Kou, No. 753, 1923, R.C. Ching" — Walker, l.c.).

General distribution: China (Nor.-West.: south. Gansu, South.-West.: nor. Sichuan).

Note. The specimens cited from Alashan with faded flowers are defective but distinguished by very tall sheaths of lower leaves, this being a characteristic of this species.

10. **G. olgae** Rgl. et Schmalh. in Izv. Obshch. lyubit. estestv., antrop. i etnogr. 34, 2 (1882) 55; Kusn. in Acta Horti Petrop. 15, 2 (1898) 309; Grossh. in Fl. SSSR, 18 (1952) 564; Fl. Kirgiz. 8 (1959) 189; Pakhom. in Opred. rast. Sr. Azii [Key to Plants of Mid. Asia] 8 (1986) 42. — *G. renardii* Rgl. in Trautv., Rgl., Maxim. et Winkl. Decas pl. nov., Petrop. (1882) 7; Kusn. in Acta Horti Petrop. 15, 2 (1898) 312. — *G. grombczewskii* Kusn. in Bull. Ac. Sci. St.-Petersb. 35 (1894) 349; id. in Acta Horti Petrop. 15, 2 (1898) 311.

— Ic.: Fl. SSSR, 18, Plate 29, fig. 1.

Described from Mid. Asia (Alay mountain range). Type in St.-Petersburg (LE).

In mountain steppes, alpine meadows, rubbly slopes and coastal pebble beds, on rocks, in spruce groves, from foothills to 3000 m.

IB. Kashgar: *West.* (Turkestania orientalis, Aug. 4, 1889 — Grombcz. typus *G. grombczewskii* Kusn. !; Kashgar, Suëk river gorge, July 27, below inflow of Suëk river into Torugart-su, July 29–1903, Lipsky, at ascent to Kurvan-kul' pass, Aug. 12, 1913 — Knorring).

IIA. Junggar: *Tien Shan* (Kapsalan mountains, Karadzhyun area, up to 2450 m, alpine meadow, July 17, 1893 — Rob.; oberstes Agias Tal, Aug. 11–20; Khaptu-su Tal bei Hauptlager, Sept. 1–5-1907, Merzb.).

General distribution: Cent. Tien Shan; Mid. Asia (West. Tien Shan, Pam.-Alay).

11. **G. olivieri** Griseb. Gen. et sp. Gentian. (1839) 278; id. in DC. Prodr. 9 (1845) 110; Boiss. Fl. or. 4 (1875) 76; Grossh. in Fl. SSSR, 18 (1952) 569; Fl. Kirgiz. 8 (1959) 190; Fl. Kazakhst. 7 (1964) 105; Pakhom. in Opred. rast. Sr. Azii [Key to Plants of Mid. Asia] 8 (1986) 42. — *G. wechniakowii* Rgl. in Acta Horti Petrop. 8, 3 (1884) 687. — *G. dahurica* auct. non Fisch.: Clarke in Hook. f. Fl. Brit. India, 4 (1883) 117, p.p.

— Ic.: Fl. Kazakhst. 7, Plate 12, fig. 5.

Described from Fore Asia (Iran). Type in Göttingen (GOET).

In sheep's fescue-wormwood and wormwood steppes in plains and mountains, on sand, in tugais and coastal meadows. Its find possible in border territories.

IA. Kashgar: *West., Nor.*

IIA. Junggar: *Jung. Alat., Tien Shan, Dzhark., Balkh.-Alak.*

General distribution: Fore Balkh., Jung. Alat., Nor. and Cent. Tien Shan; Mediterr. (east.), Asia Minor, Fore Asia, Trans-Caucasus, Mid. Asia.

64

12. G. robusta King ex Hook. f. in Hook. Ic. pl. 15 (1883) 31, tab. 1439; Clarke in Hook. f. Fl. Brit. India, 4 (1885) 734; Svenson in Brittonia, 8 (1954) 56; Hara, Chater et Williams, Enum. flow. pl. Nepal, 3 (1982) 93; Гl. Xizang, 3 (1986) 919; Fl. Sin. 62 (1988) 70. — *G. tibetica* β *robusta* (King) Kusn. in Acta Horti Petrop. 15 (1904) 323. — *G. pharica* Burk. in J. Asiat. Soc. Bengal, nov. ser. 2 (1906) 310; Fl. Xizang, l.c. 920. — *G. lhasangensis* Marq. in Kew Bull. 3 (1937) 184; Fl. Xizang, l.c. 920.

— Ic.: Fl. Xizang, 3, tab. 348, fig. 4–6; Hook. Ic. pl. tab. 1439.

Described from South. Tibet (Chumbi, on south. slope of Himalayas). Type in London (K).

On meadowy slopes of mountains, field borders, road sides, 3500–4800 m.

IIIB. Tibet: *South.* ("Khambajong, No. 298, Sept. 7, 1903 — Younghusband).

General distribution: China (South-West.: Kam), Himalayas (east.).

13. G. siphonantha Maxim. ex Kusn. in Bull. Ac. Sci. St.-Petersb. 34 (1892) 506 and in Acta Horti Petrop. 15 (1898) 316; Kanitz in Szechenyi, Wissensch, Ergebn. (1898) 717; Diels in Futterer, Durch Asien (1903) 13; Rehder in J. Arn. Arb. 14 (1933) 29; Marq in Kew Bull. 3 (1937) 166; Hao in Bot. Jahrb. 68 (1938) 629; Fl. Sin. 62 (1988) 74. — *G. siphonantha* var. *latifolia* Marq. l.c. 167; Walker in Contribs U.S. Nat. Herb. 28 (1941) 652.

— Ic.: Maxim. l.c. 1892, fig. 11–14; Fl. Sin. 62, tab. 10, fig. 8–10.

Described from Qinghai. Type in St.-Petersburg (LE). Plate IV, fig. 1.

On steppe and meadowy slopes, alpine meadows and among shrubs, coastal meadows and sandy-pebble bed shoals, 1800–4500 m.

IIIA. Qinghai: *Nanshan* (South. Tetung mountain range, alpine belt [3050 m] July 29; nor. mountain range, steppe [July 2] 1872; alpes Nan-shan, 11,000–12,000 ft. July 4 [16] 1879, No. 297 — Przew., typus !; "in ditionis Tonkerr [Donkyr] versus septentr. montis latere merid. 3000 m, July 26, 1879, Szechenyi" — Kanitz, l.c.; on Rako-gol river, 3050–3350 m, July 21, between South. Tetung mountain range and Cheibsen monastery, July 29–1880, Przew.; North. Tetung mountain range, Cherik pass, Aug. 8, 1890 — Gr.-Grzh.; Humboldt mountain range, nor. slope, 2750–3650 m, June 30; Yamatyn-umru mountains, Yamatyn-umru area, 3650 m, July 25–1894; Humboldt mountain range, nor. slope, Chonsai gorge, 2750–3350 m, July 23, 1895 — Rob.; Kuku-nor lake, Ui-yu area, Aug. 13, 1908 — Czet.; 89 km west of Xining, pass, 3450 m, Aug. 5, 1958 — Petr.; Mon'yuan', Gonshiga river valley, left tributary Peishi-khe, 3350–3720 m, Aug. 20, 1958 — Dolgushin; "an der Wasserscheide zwischen Sining-Gebiet und Kuke-nur; am Nordfuss des Semenov-Gebirges; am Kuke-nur" — Diels, l.c.; "Kokonor: am Ufer des Sees Very-dagan-tzo, 4460 m, 1930" — Hao, l.c.; "gorge Lang-tzu-tang-kou [near Sining], 2650–3100 m, 1923, R.C. Ching" — Walker, l.c.), *Amdo* ("grasslands between Labrang and Yellow River, 1927, Rock" — Rehder, l.c.).

IIIB. Tibet: *Weitzan* (valley of Alak-nor-gol river, on sand, Aug. 11, 1884 — Przew.; Burkhan-Budda mountain range, nor. slope, Khatu gorge, 4100–4400 m, July 12, 1901 — Lad.).

General distribution: China (Nor.-West.: Gansu, south-west. Nin'sya; South-West.: nor. Sichuan).

14. G. straminea Maxim. in Bull. Ac. Sci. St.-Petersb. 27 (1881) 502; Forbes et Hemsl. Index Fl. Sin. 2 (1890) 136; Kanitz in Szechenyi, Wissensch. Ergebn. (1898) 717; Diels in Futterer, Durch Asien (1903) 14; Kusn. in Acta Horti Petrop. 15 (1904) 323; Gr.-Grzh. Zap. Kitai, 3 (1907) 493; Diels in Filchner, Wissensch. Ergebn. 10, 2 (1908) 262; Rehder in J. Arn. Arb. 14 (1933) 29; Marq. in Kew Bull. 3 (1937) 167, 188; Hao in Bot. Jahrb. 68 (1938) 630; Walker in Contribs U.S. Nat. Herb. 28 (1941) 652; Hara, Chater et Williams, Enum. flow. pl. Nepal, 3 (1982) 93; Fl. Xizang, 3 (1986) 917; Fl. Sin. 62 (1988) 62.— *G. straminea* var. *humilis* Diels, 1903, l.c.

—Ic.: Fl. Xizang, 3, tab. 348, fig. 7–9; Fl. Sin. 62, tab. 9, fig. 4–8.

Described from Qinghai. Type in St.-Petersburg (LE).

In alpine meadows, meadowy steppes, shrubby associations, coastal meadows, 2000–4900 m.

IIIA. Qinghai: *Nanshan* (North. Tetung mountain range [at Chertyn-ton temple], alpine belt, Aug. 12; same site [on Gadzhur mountain], 3700 m, Aug. 14–1972, Przew., typus !; Ushilin pass, July 14, 1875—Pias.; "in ditionis Tonkerr monte 3600 m, versus septentr. Sining-fu init. VIII 1879, Szechenyi"—Kanitz, l.c.; mountain range between Nanshan and Donkyr on Rako-gol river, alpine belt, July 22; North. Tetung mountain range [on Yarlyn-gol river], alpine belt, Aug. 17–1880, Przew.; South. Kukunor mountain range, July 26; nor. bank of Kukunor, on Baga-Ulan river, Aug. 3–1890, Gr.-Grzh.; South. Kukunor mountain range, Bain-gol river, 3650 m, July 26, 1894—Rob.; vicinity of Dan'gertin town, 2450 m, Aug. 26, 1901—Lad.; "oberhalb Tankarthing bei Golien-tschuo"; "Ubergangs-pass im Süd-Kükenur Gebirge; am Nordfuss der Korallenkalkberg —Futterer"—Diels, l.c.; Loukho-Shan' mountain range, July 17, 1908; Fyi-fyi-lin pass, south-west. slope, July 20, 1909—Czet.; Chien-Shing-Cheon, south of Sining, alt. 3200 to 3300 m, No. 682, Aug. 12, 1923—R.C. Ching; "Schalakutu, 3400 m; am Fuss des ostlichen Nan-Schan, 3300 m; Sining, bei Schang-wu-chuang, 2900 m—Hao, l.c.; pass on Gulan-Lanzhou road, 2950 m, Aug. 12, 1958—Petr.; Mon'yuan', valley of Tatung-khe river near stud farm, 2800 m, Aug. 20; Mon'yuan', valley of Ganshiga river, left tributary Peishi-khe, 3350–3730 m, Aug. 20–1958, Dolgushin; pass 86 km west of Xining, Aug. 5; 108 km west of Xining and 6 km west of Daudankhe settlement, 3400 m, Aug. 5–1959, Petr.), *Amdo* (along Mudzhik river at its inflow into Huang He, July 5, 1890—Gr.-Grzh.; "Tsigi-ganba, 3400 m; Jahe-mari, 4000 m"—Hao, l.c.).

IIIB. Tibet: *Weitzan* (south. bank of Russkoe lake, July 29; Burkhan-Budda mountain range, Aug.—1884, Przew.; Russkoe lake [June 1–3, 1901]; Yantszy-tszyan river basin, Kabchzhi-kamba village, 3650 m, July 20; Nko-gun brook flowing into Rkhombo-mtso lake, 4050 m, Aug. 6–1900; Burkhan-Budda mountain range, nor. slope, Khatu gorge, 3200 m, July 2; same site, 3050–4575 m, July 12–1901, Lad.; "An'do"—Fl. Xizang, l.c.), *South.* (Gyangtse, July–Sept. 1904—Walton sub. nom. *G. waltonii;* "Shiuden Gomba, Nagong, 3900–4200 m"—Burk. l.c.).

General distribution: China (Nor.-West.; Gansu; South-West.; Sichuan, Kam; Cent · south-west. Hubei), Himalayas (east.).

Note. Judging from field labels, colour of corolla varies greatly from typical straw to pale and white, sometimes with green bands, to lilac-grey !

15. G. tianschanica Rupr. in Mem. Ac. Sci. St.-Petersb. 7 ser. 14, 4 (1869) 61; Kusn. in Acta Horti Petrop. 15 (1898) 300, incl. var. *genuina* Kusn., var. *glomerata* Kusn., var. *intermedia* Kusn., var. *roborovskii* Kusn. et

var. *pumila* Kusn.;? Hemsl. Fl. Tibet (1902) 191; Hedin, S. Tibet, 6, 3 (1922) 48; Grossh. in Fl. SSSR, 18 (1952) 563; Fl. Kirgiz. 8 (1959) 189; Fl. Kazakhst. 7 (1964) 102; Fl. Sin. 62 (1988) 66. — *G. kirilowii* Turcz. in Vestn. est. nauk Mosk. obshch. ispyt. prir. 2 (1861) 1354. descr. ross. inval.; Pakhom. in Opred. rast. Sr. Azii [Key to Plants of Mid. Asia] 8 (1986) 41. — *G. olivieri* Griseb. var. *glomerata* Rgl. in Acta Horti Petrop. 6, 2 (1880) 393. — *G. glomerata* Kusn. in Bull. Ac. Sci. St.-Petersb. 34 (1892) 507, tab. 3, fig. 10; id. in Mel. Biol. 13 (1892) 177, cum tab. fig. 24–27. — *G. regelii* Kusn. in Bull. Ac. Sci. St.-Petersb. 34 (1892) 507, id. in Mel. Biol. 13 (1892) 177, cum tab. fig. 20–23. — *G. decumbens* auct. non L. f.: Ldb. Ross. 3, 1 (1847) 64, quoad pl. alatav. p.p.; Clarke in J. Linn. Soc. Bot. 14 (1875) 440; id. in Hook. f. Fl. Brit. India, 4 (1883) 117. — *G. dahurica* auct. non Fisch.: Clarke, l.c. 117, p.p.; Persson in Bot. notis. (1938) 297. — *G. olivieri* auct. non Griseb.: Clarke, l.c. (1875) 440, p.p.

—Ic.: Gartenfl. 1882, tab. 1069; Bull. Ac. Sci. St.-Petersb. 34, 2, tab. 3, fig. 10; Fl. Kazakhst. 7, Plate 13, fig. 8.

Described from Cent. Tien Shan. Type in St.-Petersburg (LE).

On wet rubbly-rocky and meadowy slopes, scrubs, spruce groves and among juniper beds in middle and upper belts of mountains, 1500–3600 m.

60    IB. Kashgar: *West.* (Tokhtakhon mountains, nor. slope, 3050–3650 m, in meadows, July 21, 1889 — Rob. [var. *roborowskii* Kusn. l.c.]).

IIA. Junggar: *Jung Alat.* (Jung. Alat. [in upper Borotala], 2150–2450 m, Aug. 5, 1878 — A. Reg.; Borotala basin, below Koketau pass, July 21, 1909 — Lipsky; Arba-Kezen' mountains in Toli area, No. 2547, Aug. 6, 1957 — Kuan; Barlyk mountain range, No. 1008, Aug. 6, in Toli town region, Barktok — Arba-Kezen', No. 1313, Aug. 7; same site, 1780 m, No. 1366, Aug. 8–1957, Shen; south-west. fringe of Maili-Barlyk mountain range at Karaganda pass on road to "Junggar gate" from Toli, nor. slope, Aug. 15, 1957 — Yun, S.-y. Lee, Yuan'; Toli dist., around Ebi-nur lake, 2020 m, No. 1863, Aug. 17; 25 km west of Ven'tsyuan', Intermontane plain, 2020 m, No. 2007, Aug. 24; 10 km east of Ven'tsyuan', 2330 m, in forest, No. 2088, Aug. 26–1957, Shen), *Tien Shan* (Talki river, July 9; vicinity of Suidun, July 16; Talkinskoe gorge, July 17; Sairam lake, July 20; Dzhagastai, 1500–2150 m, Aug. 7; Muzart picket in Tekes valley, 1500 m, Aug. 15–16; Tekes valley in midcourse, 2150–2450 m, Aug. 17–1877, A. Reg.; Maralty, on Muzart gorge, 1850 m, Aug. 1; Urten-Muzart gorge, Aug. 4; in Narynkol valley, Aug. 8–1877, Fet.; Talki gorge, 2000 m, July 10; Urtak-sary river at Sairam lake, July 18–20; Sairam lake, July 20–1878, Fet.; Dzhirgalan [upper Khorgos] 1500–1850 m, July 26; Kok-kamyr mountains, 2150–2750 m, July 27; Kyzemchek, 2150–2750 m, July 29; in Kok-kamyr mountains, 2150–2450 m, July 31; Chubaty pass, 2450–2750 m, Aug. 2; upper Borotala, 2600 m, Aug. 10; south-east. bank of Sairam lake, Aug. 29; Chishkan-toka on Kash river, 1500 m, Sept. 11–1878, A. Reg.; Toguz-torau valley, Aug. 16 and 18, 1878 — Fet.; Aryslyn, 2450–2750 m, July 10; Turgun-tsagan,? July–1879, A. Reg.; on Tekes river, 3350 m, July 12; B. Yuldus, 2450–2750 m, Aug. 5–1893, Rob.; Sarydschass, zwischen Lager und Gletscherende, Aug. 2–8, 1903; mitleres und unteres Agias Tal, Aug. 11–20; beim Hauptlager im Tal Khaptu-su, Sept. 1–5–1907, Merzb.; "Köl, about 2700 m, Aug. 2, 1932" — Persson, l.c.; Talkhin mountain range foothills on south-east. bank of Sairam lake, 2150 m, plain, Aug. 18, 1953 — Mois.; 3 km south of Tekes town, 1790 m, No. 657, Aug. 7; 8 km east of Chzhaosu town, in intermontane basin, 1740 m, No. 769, Aug. 11;

vicinity of Syat settlement, 2650 m, No. 1331, Aug. 11; east of Aksu on road to Chzhaosu, No. 1592, Aug. 16–1957, Shen; in Dzhagastai region, 1800 m, No. 3194, Aug. 8; stud farm around Chzhaosu town, intermontane basin, No. 3258, Aug. 11; between Gunlyu and Shakhe, No. 3651, Aug. 18–1957, Kuan; upper Borotala river 7–8 km nor.-west of Shivutin-daba pass on road to Sairamnur from Arasan, Aug. 18, 1957 — Yun., S.-y. Lee, Yuan'; Bai dist., Oi-terek area, 2900 m, No. 8234, Sept. 7; Bai dist., valley of Tukbel'chi, 2500 m, on terrace, No. 8377, Sept. 9–1958, Lee et Chu; valley of Muzart river in upper Tunu-daban near Oi-terek area, 2900 m, spruce forest, Sept. 7, 1958 — Yun et Yuan').

IIIB. Tibet: *Chang Tang* ("Tsaidam, sandy soil near water at 15,400 ft [4700 m] 1891, Thorold" — Hemsl. l.c.?; "Sarik-kol, 3469 m, Aug. 5, 1896" — Hedin, l.c.).

General distribution: Jung.-Tarb., Nor. and Cent. Tien Shan; Mid. Asia (West. Tien Shan, Pam.-Alay), Himalayas (west., Kashmir).

Note. The species is very close to *G. decumbens* L.f., from which it differs foremost in developed, 5–7 mm long, linear teeth, generally entire calyx and very narrow tubular-infundibular corolla. Calyx is, however, quite often split up to half. Flowers either on long pedicels, in loose racemose inflorescence or subsessile, aggregated in umbelliform or semicapitate inflorescence, sometimes arranged stepwise, rarely single. Occasionally, plants are found with light-blue — light-yellow corolla or albinic corolla.

16. G. tibetica King ex Hook. f. in Hook. Ic. pl. 15 (1883) 33, tab. 1441; Clarke in Hook. f. Fl. Brit. India, 4 (1885) 733; Kusn. in Acta Horti Petrop. 15 (1904) 322; Hara, Chater et Williams, Enum. flow. pl. Nepal, 3 (1982) 94; Fl. Xizang, 3 (1986) 917; Fl. Sin. 62 (1988) 68. — *G. brevidens* Rgl. in Acta Horti Petrop. 10 (1887) 376.

— Ic.: Hook. Ic. pl. 15, tab. 1441; Curtis' Bot. Mag. 123, pl. 7528; Fl. Xizang, 3, tab. 348, fig. 1–3.

61     Described from South. Tibet (Chumbi on south. slope of Himalayas). Types in London (K).

Among shrubs, along fringes of farms, 2100–4200 m.

IIIB. Tibet: *South.?*

General distribution: China (South-West.: Kam), Himalayas (east.).

Note. Found in the border regions of South. Tibet along south. slope of Himalayas and in south-west. Sikang basin but has not been reported so far within the region.

17. G. waltonii Burk. in J. Asiat. Soc. Bengal, nov. ser. 2 (1936) 310; Marq. in Kew Bull. 3 (1937) 188; Fl. Xizang, 3 (1986) 915; Fl. Sin. 62 (1988) 59.

— Ic.: Fl. Sin. 62, tab. 9, fig. 1–3.

Described from South. Tibet. Type in London (K).

Along turf-covered meadowy and rubbly-rocky slopes of mounds and mountains, on rocks, 3000–4800 m.

IIIB. Tibet: *South.* ("in valle rivuli Kyi-chu dicti, prope Lhasa, No. 1645, Walton — typus !; Lhasa, 12,000 ft. Walton, Waddell and at Gyangtze, No. 1648, Walton" — Burk. l.c.; "Gyantse, rocky hillside, 4200 m, Sept. 15, 1933; Pomo Tso, among stones, 4800 m, Sept. 11, 1933 — Ludlow et Sherriff" — Marq. l.c.; "Lhasa, Shigatsze, Mochzhugunka" — Fl. Xizang, l.c.).

General distribution: China (South-West.: Kam).

18. G. walujewii Rgl. et Schmalh. in Acta Horti Petrop. 6 (1879) 334; Kusn. ibid. 15, 2 (1898) 313; Persson in Bot. notis. (1938) 297; Grossh. in Fl. SSSR, 18 (1952) 565; Fl. Kazakhst. 7 (1964) 103; Fl. Sin. 62 (1988) 70. — *G. kesselringii* Rgl. in Acta Horti Petrop. 7, 2 (1881) 548. — *G. walujewii* β *kesselringii* Kusn. in Acta Horti Petrop. 15, 2 (1898) 314.

— Ic.: Gartenflora, tab. 1087, fig. 3–4; Fl. SSSR, 18, Plate 29, fig. 2; Fl. Kazakhst. 7, plate 11, fig. 9.

Described from East. Tien Shan. Type in St.-Petersburg (LE).

On steppe and meadowy slopes, coastal meadows and pebble beds, 1500–3000 m.

IIA. Junggar: *Tien Shan* (Nanshan-kou pass, Aug. 31, 1875 — Pias.; Dzhirgalan, 1500–1850 m, July 26; same site, 2450–2750 m, July 26; south. fringe of Kokkamyr plateau, 1850 m, July 26; Kokkamyr mountains, Bogdo mountain, 2450–2750 m, July 27; in Kokkamyr mountains, 3050 m, July 27–1878. A. Reg.; in Yuldus mountains, Sept. 1878 — Fet., typus !; Nilki [on Kash river], 2150 m, June 30; Turgun-Tsagan [on Kash river], July 3; left bank of Kash river, 2750 m, July 15; Aryslyn foothills, 1850–2450 m, July 19 — typus *G. kesselringii* Rgl.; Aryslyn, 2450–2750 m, July 19 — Aryslyn estuary on Kash river, 1650 m, July 22; between Saryk and Mengute, 1850–2150 m, July 27; Mengute, on nor. slope of Irenkhabirga, 2750 m, Aug. 2; lower Aryslyn, 1850 m, Aug. 20; on Kunges river, 2750 m, Aug. 31–1879, A. Reg.; Khaidyk-gol, 2750 m, steppe, Aug. 10, 1893 — Rob.; am Wege von Fucan zur Bogdo-ola, in der Lehm- und Kies-steppe, Aug. 2–3, 1908 — Merzb.; "Kunges, about 1600 m, Aug. 24, 1932" — Persson, l.c.; Ulyastai, 1750 m, No. 1216, July 26, 1957 — Shen; in Mal. Yuldus, 2250 m, on alluvium, No. 6376, Aug. 2, 1958 — Lee et Chu, south. slope of Narat mountain range in Bol. Yuldus basin, 10–12 km nor.-west of Bain-Bulak settlement, Aug. 8, 1958 — Yun. et Yuan').

General distribution: Jung.-Tarb.

## Section 2. Frigida Kusn.,
## Subsection Monopodiae (H. Smith) T.N. Ho

19. G. callistantha Diels et Gilg in Futterer, Durch Asien, Bot. repr. 3 (1903) 14; Fl. Sin. 62 (1988) 78. — *G. szechenyi* auct. non Kanitz: H. Smith in Hand.-Mazz. Symb. Sin. 7 (1936) 975, pro syn.; Marq. in Kew Bull. 3 (1937) 161, pro syn.

— Ic.: Futterer, l.c. tab. 1 A; Fl. Sin. 62, tab. 11, fig. 3–4.

Described from Qinghai. Type in Berlin (B).

In steppe and alpine meadows, on talus along sunny slopes, 3400–4900 m.

IIIA. Qinghai: *Amdo* ("Semenow-Gebirge bei Lager XVI, bei 4000 m, No. 145, 153, Sept. 1898, Futterer", typus ! — Diels, l.c.).

IIIB. Tibet: *Weitzan* ("nor.-east. Tibet" — Fl. Sin. l.c.).

General distribution: China (Nor.-West.: Gansu, south-west.).

20. G. dolichocalyx T.N. Ho in Acta phytotax. sin. 23, 1 (1985) 43; Fl. Sin. 62 (1988) 86.

—Ic.: Acta phytotax. sin. 23, 1, tab. 1, fig. 5–7 (sub nom. *G. longicalyx* T.N. Ho).

Described from South-West. China (Sichuan). Type in Chongtu (SZ).

In alpine meadows, among shrubby associations, roadsides, 3000–3800 m.

IIIB. Tibet: *Weitzan* ("Tszyuchzhi").

General distribution: China (Nor.-West.: south-west. Gansu; South-West.: nor.-west. Sichuan).

21. G. farreri Balf. f. in Trans. a. Proc. Bot. Soc. Edinb. 27 (1918) 248; Marq. in Kew Bull. 3 (1937) 159; Fl. Xizang, 3 (1986) 938; Fl. Sin. 62 (1988) 95.

—Ic: Fl. Xizang, 3, tab. 359; Fl. Sin. 62, tab. 14, fig. 4–6.

Described from Nor.-West. China (Gansu). Type in Edinburgh (E). Plate III, fig. 4.

On meadowy slopes, alpine grasslands, coastal meadows, wet and marshy sites, 2400–4600 m.

IIIA. Qinghai: *Nanshan* (south. descent of Bukhyin-daban mountain range in upper Ara-gol, 4250 m, Sept. 12, 1894—Rob.; Kuku-nor lake, valley on south. bank, 3080 m, Aug. 20; South. Kukunor mountain range, nor. slope, 3100–3350 m, Aug. 20; Kuku-nor lake, south. bank, near Nara-saren-khutul' pass, 3200 m, near Tsagan-yan'pin town ruins, Aug. 21; Choibsen-khit temple, 2450 m along mountain descents and along valley, Aug. 30–1901, Lad.; Kuku-nor lake, Ui-yu area, lower and middle belts, along southern slope, Aug. 12; South. Kuku-nor mountain range, midbelt, Aug. 16–1908, Czet.; pass on highway to Lanzhou from Uvei, 2800 m, Oct. 9, 1957—Yun., S-y. Lee, Yuan').

IIIB. Tibet: *Weitzan* (Rkhombo-mtso lake, 3950 m, on meadowy descents of mountains and along lake banks, Aug. 4; same site, vicinity of lake, 4000 m, Aug. 6; vicinity of Chzherku monastery, 3470 m, along mountain descents, Aug. 9–1900, Lad.), South (Chumbu-la, 10–15 miles north of Lhasa, Sept. 1904—Walton; "Lhasa, Shigatsze, Linzhou, Leinyaotszy"—Fl. Xizang, l.c.).

General distribution: China (Nor.-West.: Gansu; South-West.: Sichuan).

Note. Colour of corolla varies from light blue to dark blue and lilac.

22. G. futtereri Diels et Gilg in Futterer, Durch Asien, Bot. repr. 3 (1903) 14; Rehder in J. Arn. Arb. 14 (1933) 28; Marq. in Kew Bull. 3 (1937) 159; Fl. Xizang, 3 (1986) 940; Fl. Sin. 62 (1988) 96.— *G. sino-ornata* auct. non Balf. f.: Hao in Bot. Jahrb. 68 (1938) 629.

—Ic.: Futterer, l.c. tab. I B; Fl. Sin. 62, tab. 14, fig. 7–9.

Described from Tibet. Type in Berlin (B).

In moist and swampy meadows, swamps, along banks of rivers.

IIIA. Qinghai: *Nanshan* ("Kokonor region, 1925, Rock"—Rehder, l.c.; "Kokonor: Amne-Matchin, 4500 bis 5000 m; auf dem südabhängen des Gebirges Selgen, um 3800 m; am Füsse des Gebirges zwischen dem See Kokonor und der Stadt Tenkar; auf dem Gebirge Mingge, 3900 m—1930"—Hao, l.c.).

63  IIIB. Tibet: *Weitzan* ("östlich des Hoang-ho am Nordfuss des Dschupar-Gebirges, No. 195, Sept. 29; typus !; auf den Bergen südlich vom Baa-Flusse, No. 200, Nov. 4, Futterer"—Diels, l.c.).

General distribution: China (South-West.).

23. G. hexaphylla Maxim. ex Kusn. in Bull. Ac. Sci. St.-Petersb. 35 (1894) 349; Kusn. in Acta Horti Petrop. 15, 2 (1898) 270; H. Smith in Hand.-Mazz. Symb. Sin. 7 (1936) 974; Marq. in Kew Bull. 3 (1937) 158; Fl. Sin. 62 (1988) 84.

—Ic.: Maxim. l.c. fig. 31–34; Fl. Sin. 62, tab. 12, fig. 6–9.

Described from South-West. China (Sichuan). Type in St.-Petersburg (LE).

On meadowy slopes of mountains, alpine meadows, wet scrubs, roadsides, up to 4400 m.

IIIA. Qinghai: *Amdo* ("south-east. Qinghai" — Fl. Sin. l.c.).

General distribution: China (Nor.-West.: south. Gansu, South-West.: nor.-west. Sichuan).

24. G. stipitata Edgew. in Trans. Linn. Soc. 20 (1846) 84; H. Smith in Kew Bull. 15, 1 (1961) 50; Hara, Chater et Williams, Enum. flow. Pl. Nepal, 3 (1982) 93; Fl. Sin. 62 (1988) 75. — *G. tizuensis* Franch. in Bull. Soc. Bot. France, 43 (1896) 494; Marq. in Kew Bull. 3 (1937) 161; H. Smith, l.c. 50; Fl. Xizang, 3 (1986) 934. — *G. cachemirica* auct. non Decne: Clarke in Hook. f. Fl. Brit. India, 4 (1883) 115. — *G. depressa* auct. non D. Don: Kusn. in Acta Horti Petrop. 15, 3 (1904) 342, p.p.

—Ic.: Kew Bull. 15, 1, tab. 2, fig. 1–2; Fl. Sin. 62, tab. 11, fig. 1–2; Fl. Xizang, 3, tab. 357 (sub. nom. *G. tizuensis* Franch.).

Described from Himalayas. Type in london (BM?).

In alpine meadows, along banks of brooks and in coastal meadows, wet talus, 3200-4600 m.

IIIA. Qinghai (Fl. sin. l.c.).

General distribution: China (South-West.), Himalayas.

25. G. szechenyi Kanitz, Pl. exped. B. Szechenyi in Asia centr. collet. enum. (1891) 40 and in B. Szechenyi, Wissensch. Ergebn. 2 (1898) 717; Kusn. in Acta Horti Petrop. 15 (1898) 267; Marq. in Kew Bull. 3 (1937) 186, excl. sin. *G. callistantha* Diels et Gilg. — *G. georgei* Diels in Notes Bot. Gard. Edinb. 5 (1912) 221; Marq. in Kew Bull. 3 (1937) 161; Hao in Bot. Jahrb. 68 (1938) 629; Fl. Xizang, 3 (1986) 934; Fl. Sin. 62 (1988) 79.

—Ic.: Kanitz, l.c. (1898) tab. 4, fig. 2; Fl. Sin. 62, tab. 11, fig. 5–7.

Described from South-West. China (Sichuan). Type in Wien (Vienna) (W) ?

On meadowy slopes, coastal meadows, 3900–4800 m.

IIIA. Qinghai: *Amdo* ("Kokonor: Jahe-mari, 4000 m; Amne-Matchin, 4500 m, 1930" — Hao, l.c.).

IIIB. Tibet: *Weitzan* (Yangtze river, along Go-chyu brook, 3950 m, on meadowy descents of mountains and along banks of rivers and brooks, Aug. 23, 1900 — Lad.; "Bizhu" — Fl. Xizang, l.c.).

General distribution: China (South-West.: west. Sichuan, nor.-west. Yunnan, Kam).

26. G. veitchiorum Hemsl. in Gard. Chron. 46 (1909) 178; H. Smith in Hand.-Mazz. Symb. Sin. 7 (1936) 972; Marq. in Kew Bull. 3 (1937) 159, 186, excl. var. *caelestis* Marq.; Fl. Xizang, 3 (1986) 936; Fl. Sin. 62 (1988) 90.
— *G. ornata* var. *obtusiloba* Franch. in Bull. Soc. Bot. France, 43 (1896) 493.
— *G. ornata* var. *acutiloba* Franch. l.c. 494.

—Ic.: Gard. Chron. 46, tab. 74; Fl. Sin. 62, tab. 13, fig. 3–5.

Described from South-West. China (Sichuan). Type in London (K).

In alpine meadows and meadowy slopes, shoals, coastal scrubs, 3000–4700 m.

IIIA. Qinghai ("Qinghai"—Fl. Sin. l.c.).
IIIB. Tibet: *South.* ("Linzhou, Nan'mulin—Fl. Xizang, l.c.).
General distribution: China (Nor.-West.: Gansu, South-West.).

## Section 2. Frigida Kusn.
## subsection Sympodia H. Smith

27. G. algida Pall. Fl. Ross. 1, 2 (1788) 107; Kusn. in Acta Horti Petrop. 15 (1898) 259; Danguy in Bull. Mus. nat. hist. natur. 20 (1914) 75, cum auct. Haenke; Persson in Bot. notis. (1938) 297; Kryl. Fl. Zap. Sib. 9 (1937) 2183; Kitag. Lin. Fl. Mansh. (1939) 358; Grossh. in Fl. SSSR, 18 (1952) 559; Grub. Konsp. fl. MNR [Conspectus of Flora of Mongolian People's Republic] (1955) 222; Fl. Kirgiz. 8 (1959) 188; Fl. Kazakhst. 7 (1964) 100; Grub. Opred. rast. Mong. (Key to Plants of Mongolia] (1982) 201; Pakhom. in Opred. rast. Sr. Azii [Key to Plants of Mid. Asia] 8 (1986) 40; Fl. Sin. 62 (1988) 109.— *G. frigida* γ *algida* Ldb. Fl. Ross. 3 (1847) 65.

—Ic.: Pall. l.c. tab. 95; Gartenfl. 29, tab. 1006; Fl. Kazakhst. 7, Plate 11, fig. 7.

Described from East. Siberia. Type in London (BM) or Berlin (B).

In alpine meadows, *Cobresia* groves, swamps, syrts—watershed uplands, standing moraines, along wet rubbly-rocky slopes, along banks of brooks and rivers, larch forests and groves, in upper belt of mountains, 2500–3600 m.

IA. Mongolia: *Khobd.* (Zwischen 1 [Kaak], 2 und 3 Piketen, June 8–July 12, 1870—Kalning; south-east. spur of Turgen' mountain range, Kashgurt mountain, south. slope. July 21; Umne-Otor-ula mountains 30 km nor.-east of Kobdo somon, 2750 m. July 30–1977. Karam., Sanczir et al), *Mong. Alt* ("entre Oulioungour et Kobdo, alt. 2560 m, Altai, Sept. 7, 1895—Chaff."—Danguy, l.c.; mountain slope on Ulyasty river upper courses, July 22; mountain slope on Naryn river upper courses, July 23–1898, Klem.; upper Dzhirgalanta river, Bulugun tributary, July 25, 1906—Sap.; Kharagaitu-daba pass in upper Indertiin-gol, July 24; Bulugun river basin, upper Ketsu-Sairin-gol along moraine and slopes toward glacier, July 26–1947, Yun.), *Gobi Alt.* (Ikhe-Bogdo-ula).

IIA. Junggar: *Cis-Alt.* (mountains at source of Kandagatai river, Sept. 18; Oi-chilik, alpine meadow, Sept. 20–1876, Pot.; Qinhe dist., Dakheitsza, 2600 m, No. 1707, Aug. 9, 1956—Ching), *Jung. Alat.* (westward of Ven'tsyuan', waterdivide, 3000 m, No. 2023, Aug. 25, 1957—Shen; in Ven'tsyuan' dist., 2940 m, No. 4639, Aug. 25, 1957—Kuan), *Tien Shan* (Itsym-Bukhty [Kul'dzha region], June 22, 1875—Larionov; Urten-Muzart pass, Aug. 3; Urten-Muzart gorge, Aug. 4–1877, Fet.: Muzart pass, 3050–3500 m, 'on south. slope above large glacier, Aug. 18; upper Muzart below pass, 2750–3200 m; Aug. 19–1877, A. Reg.; [nor.-west of Kul'dzha] July 4; in Kul'dzha region, July 9; Sairam lake, July 18–1878, Fet.; upper Borotala, Kazan pass, 2750–3050 m, Aug. 10; Kazan lake [Khorgos]? Aug.—1878; lower Arystyn, 2150–2450 m, July 20; Mengute [Irenkhabirga], 3050–3350 m, July 4 and Aug. 2; Kash river valley, south. slope, Aug. 17–18–1879, A. Reg.; Kapsalan mountains, Karadzhyun area, up to 2450 m, July 17; near Kapsalan pass, 2450–3050 m, July 18–1893, Rob.; Sarydschass, Aug. 2–8, 1903; Passe zwischen Kinsu und Kurdai, July 3, 1907; Oberstes Agias Tal, Aug. 11, 20, 1907; Lager am Südrande des Bogdo-ola, Aug. 26–29, 1908—Merzb.; "Tilemet pass, about 3200 m, Aug. 8, 1932"—Persson, l.c.; 3 km south of Danu, 3150 m, No. 417, July 18, 1957—Shen; Danu-daban pass, between upper courses of Ulan-usu river and Danu-gol, 3600 m, nival (snow) belt, July 19, 1957—Yun., S.-i. Lee, Yuan'; along Danu river, 3000 m, on slope, No. 2137 and same site, 2700 m, No. 2074, July 21, 1957—Kuan; 7–8 km south of Danu, 3450 m, on south. slope, No. 510, July 22, 1957—Shen; 10 km nor. of Chzhaosu, at summit, No. 3300, Aug. 14; Kentelek mountains, in Aksu region, 3000 m, No. 3517, Aug. 14–1957, Kuan; 70 km south of Ven'tsyuan', 2810 m, No. 1476, Aug. 14; 20 km south of Ven'tsyuan', 2900–3200 m, No. 1547, Aug. 14; Telisygai? No. 905, Aug. 15; 3 km south of Yakou, Barchat, 2900 m, No. 1673 and No. 1703, Aug. 30–1957, Shen; Nilki to Dzinkho through Yakou, on slope, No. 4038, Sept. 1, 1957—Kuan; Kotel' pass on road to Yuldus from Karashar, 3100–3200 m, Aug. 1; Mal. Yuldus basin, 1–2 km west of Kotel' pass, alpine belt, Aug. 15–1958, Yun. et Yuan'; Khetszi district, 2900 m, No. 7091, Aug. 5; in Khetszi district at Tsagan-nur pass, 3100 m, No. 6533, Aug. 15–1958, Lee et Chu).

General distribution: Jung.-Tarb., Nor. and Cent. Tien Shan; West. Sib. (south. Altay), East. Sib. (south.), Nor. Mong. (Fore Hubs., Hent., Hang.), China (Altay, Dunbei: Chanbaishan'), Japan.

28. G. nubigena Edgew. in Trans. Linn. Soc. (Bot.) 20 (1846) 85; Clarke in Hook, f. Fl. Brit. India, 4 (1883) 116; Hemsl. Fl. Tibet (1902) 191; Hedin, S. Tibet, 6, 3 (1922) 47; Fl. Xizang, 3 (1986) 925; Fl. Sin. 62 (1988) 112, p.p.— *G. algida* ε *nubigena* (Edgew.) Kusn. in Acta Horti Petrop. 15, 2 (1898) 226; Hara, Chater et Williams, Enum. flow. pl. Nepal, 3 (1982) 91.

—Ic.: Edgew. l.c. fig. 49; Fl. Sin. 62, tab. 18, fig. 1–3.

Described from West. Himalayas. Type in London (BM?).

In alpine meadows, moraines, coastal meadows, 4000–5300 m.

IIIB. Tibet: *Weitzan* (Burkhan-Budda mountain range, alpine belt, 4300–4800 m, in meadows, Aug. 12, 1884—Przew.; Yantszy-tszyan basin: Chamudug-la pass, 4800 m, along nor. and south. descents of pass and at its peak, July 26; along valley of Nkogun brook flowing into Rkhombo-mtso lake, 4050 m, Aug. 6–1900, Lad.; "Bizhu"—Fl. Xizang, l.c.), *South.* ("Balch pass and Rakas Tal, 15,000–17,000 ft., 1848, Strachey et Winterbottom"—Hemsl. l.c.; "valley of Upper Tsangpo on road between Dongbo, 4598 m and Tuksum, 4596 m, July 1, 1907"—Hedin, l.c.; "Nimu"—Fl. Xizang, l.c.).

General distribution: China (Nor.-West.: Gansu, South-West.: Sichuan), Himalayas.

29. G. przewalskii Maxim. in Bull. Ac. Sci. St.-Petersb. 27 (1881) 502; Forbes et Hemsl. Index Fl. Sin. 2 (1890) 132; Marq. in Kew Bull. 3 (1937) 164; Hao in Bot. Jahrb. 68 (1938) 629; Walker in Contribs U.S. Nat. Herb. 28, 4 (1941) 652; Fl. Xizang, 3 (1986) 351. — *G. algida* δ *przewalskii* (Maxim.) Kusn. in Acta Horti Petrop. 15, 2 (1898) 265; Hara, Chater et Williams, Enum. flow. pl. Nepal, 3 (1982) 91. — *G. chingii* Marq. in Kew Bull. (1931) 83 and 3 (1937) 164; Walker, l.c. 651; Fl. Xizang, 3 (1986) 923. — *G. nubigena* auct. Fl. Sin. 62 (1988) 112, p.p.

—Ic.: Fl. Xizang, 3, tab. 351.

Described from Qinghai. Type in St.-Petersburg (LE). Plate III, fig. 1.

In alpine meadows, moraines, rocky placers and talus, coastal meadows and shoals, 3000–5600 m.

IIIA. Qinghai: *Nanshan* (Regio alpina jugi N. a fl. Tetung in pratis alpinis frequens, Aug. 10 [22] 1872 — Przew., typus!; North. Tetung mountain range, alpine belt, Aug. 17, 1880 — Przew.; valley of Babo-kho river, 3350 m, meadow, Aug. 9, 1890 — Gr.-Grzh.; South. Kukunor mountain range, nor. slope, 3950–4250 m, Naion-khutul' pass, south. slope of Bukhyn-daban up to 4350 m and ridges in upper Bukhaina river, Aug. 1, 1894 — Rob.; Mon'yuan', at head of Ganshiga river, Peishi-khe tributary, 3900–4300 m, glacial moraine, Aug. 18, 1958 — Dolgushin; "Koko-nor auf dem Selgen, 4800 m, 1930" — Hao, l.c.).

IIIB. Tibet: *Weitzan* (south. bank of Russkoe lake, 4100 m, July 31, 1884 — Przew.; Burkhan-Budda mountain range, nor. slope, Khatu gorge, 4575 m, under Khatu-daba pass; in wet placers, July 12; same site, Khatu gorge, 4250 m, in alpine belt, July 18–1901, Lad.), South. ("Chzhunba, Pulan'" — Fl. Xizang, l.c.).

General distribution: China (Nor.-West.: Gansu, South-West.: west. Sichuan).

Note. Colour of corolla varies from dark-blue–violet to light blue.

30. G. purdomii Marq. in Kew Bull. (1928) 55 and 3 (1937) 164; Fl. Xizang, 3 (1986) 927; Fl. Sin. 62 (1988) 110. — *G. algida* auct. non Pall.: Rehder in J. Arn. Arb. 14 (1933) 28; Hao in Bot. Jahrb. 68 (1938) 628. — *G. frigida* auct. non Haenke: Danguy in Bull. Mus. nat. hist. natur. 17, 5 (1911) 10.

—Ic.: Fl. Xizang, 3, tab. 353; Fl. Sin. 62, tab. 17, fig. 1–4 (sub. *G. nubigena*!).

Described from Nor.-West. China (Gansu). Type in London (K).

In alpine meadows, shifting rocky talus and placers, wet meadowy slopes and rocks, 2700–4800 m.

IIIA. Qinghai: *Nanshan* ("Rochers du col de Tapan-Chan, alt. 4000 m, July 10, 1908 — Vaillant" — Danguy, l.c.), *Amdo* ("Kokonor; auf dem Gebirge Mingga, südlich des klosters Taschin-sze, 3500–4500 m, Anguma auf den Abhängen, 4500 m, 1930" — Hao, l.c.; "grasslands between Labrang and Yellow River — Rock" — Rehder, l.c.).

IIIB. Tibet: *Weitzan* ("Bizhu" — Fl. Xizang, l.c.).

General distribution: China (Nor.-West: Gansu, South-West.: west. Sichuan).

## Section 2. Frigida Kusn.
## subsection Annua (Marq. 1937) Grub. comb. nova
## (section *Microspermae* T.N. Ho)

31. G. tongolensis Franch. in Bull. Soc. Bot. France, 43 (1896) 490; H. Smith in Hand.-Mazz. Symb. Sin 7 (1936) 978; Marq. in Kew Bull. 3 (1937) 160; Fl. Xizang, 3 (1986) 944; Fl. Sin. 62 (1988) 142.

—Ic.: Fl. Xizang, 3, tab. 360; Fl. Sin. 62, tab. 22, fig. 7–9.

Described from South-West. China (Sichuan). Type in Paris (P).

In alpine meadows, along sides of mountain raods, 3700–4800 m.

IIIB. Tibet: *Weitzan* ("Bizhu" — Fl. Xizang, l.c.).

General distribution: China (South-West).

32. G. vernayi Marq. in Hook. Ic. Pl. 34 (1937) tab. 3330, fig. 8–13; Hara, Chater et Williams, Enum. flow. pl. Nepal, 3 (1982) 94; Fl. Xizang, 3 (1986) 942; Fl. Sin. 62 (1988) 142.

—Ic.: Hook. Ic. Pl. 34, tab. 3330, fig. 8–13.

Described from South. Tibet. Type in London (K).

In alpine meadows and meadowy slopes, 4200–5000 m.

IIIB. Tibet: *South.* ("Lhasa, Lankatsza, Nan'mulin" — fl. Xizang, l.c.).

General distribution: China (South-West.: Kam), Himalayas (east.).

## Section 3. Pneumonantha Gaudin

33. G. dschungarica Rgl. in Acta Horti Petrop. 6 (1880) 334; Kusn. ibid, 15, 2 (1898) 236; Grossh. in Fl. SSSR, 18 (1952) 550; Fl. Kazakhst. 7 (1964) 100; Pakhom. in Opred. rast. Sr. Azii [Key to Plants of Mid. Asia] 8 (1986) 39.

—Ic.: Fl. Kazakhst. 7, Plate 11, fig. 5.

Described from Sinkiang (Jung. Ala Tau). Type in St.-Petersburg (LE).

On meadowy slopes and in coastal meadows in midbelt of mountains.

IIA. Junggar: *Jung. Alat.* (Dschungarischer Alatau, gegen die Borotala, 7000–8000 ft, Aug. 5 [17] 1878, A. Reg.—typus !).

General distribution: Jung.-Tarb. (Jung. Ala Tau).

34. G. fischeri P. Smirn. in Bull. Soc. Natur. Moscou, 45, 2 (1937) 95; in Pl. alt. exic. (1937) No. 66; Grossh. in Fl. SSSR, 18 (1952) 549; Fl. Kazakhst. 7 (1964) 99; Pakhom. in Opred. rast. Sr. Azii [Key to Plants of Mid. Asia] 8 (1986) 39. — *G. septemfida* auct. non Pall.: Ldb. Fl. Ross. 3, 1 (1847) 67; Kryl. Fl. Zap. Sib. 9 (1937) 2182. — *G. septemfida* α *genuina* Kuzn. in Acta Horti Petrop. 15, 2 (1898) 239, 240. — *G. gebleri* Fisch. ex Bge. non Ldb. in Nouv. Mem. Soc. natur. Moscou, 1 (1829), 218, nomen, non Ldb. ibid, 211.

67     Described from Altay (Naryn mountain range). Type in Moscow (MW). Plate III. fig. 5.

In forest meadows and glades, coastal meadows, alpine tundra.

IIA. Junggar: *Jung. Alat.* (nor. slope against Sary-agach brook at alt. 2450–2750 m in coniferous forest, Aug.; opposite Borotala, 2150–2450 m, Aug.—1878, A. Reg.).

General distribution: Jung.-Tarb.; West. Sib. (south. Altay).

## Section 4. Isomeria Kusn.

35. G. amplicrater Burk. in J. Asiat. Soc. Bengal, n.s. 2 (1906) 312; H. Smith in Kew Bull. 15, 1 (1961) 51; Hara, Chater et Williams, Enum. flow. pl. Nepal, 3 (1982) 91; Fl. Xizang, 3 (1986) 930; Fl. Sin. 62 (1988) 126.

—Ic.: Fl. Xizang, 3, tab. 355.

Described from Tibet. Type in London (K).

In swampy meadows, along wet ravines of seasonal streams, 4000–4400 m.

IIIB. Tibet: *South.* ("Prope Lhasa ad fauces Pembu-la dictas, Walton, No. 1657", typus !—Burk, l.c.; "Nimu, Linzhou"—Fl. Xizang, l.c.).

General distribution: China (South-West.: Kam), Himalayas (east.).

36. G. urnula H. Smith in Kew Bull. 15, 1 (1961) 51; Hara, Chater et Williams, Enum. flow. pl. Nepal, 3 (1982) 94; Fl. Xizang, 3 (1986) 932; Fl. Sin. 62 (1988) 128.

—Ic.: Fl. Sin. 62, tab. 20, fig. 4–6.

Described from Himalayas (Bhutan). Type in London (BM).

In alpine meadows, talus, sandy-rubbly slopes, 3700–5200 m.

IIIA. Qinghai: *Amdo* ("south-west. Qinghai"—Fl. Sin. l.c.).

IIIB. Tibet: *Weitzan* ("Bizhu"—Fl. Xizang, l.c.), *South.* ("Batsin, Linzhou, Lhasa, Nan'mulin"—fl. Xizang, l.c.).

General distribution: Himalayas (east.).

## Section 5. Stenogyne Franch.

37. G. striata Maxim. in Bull. Ac. Sci. St.-Petersb. 27 (1881) 501; Forbes et Hemsl. Index Fl. Sin. 2 (1890) 136; Kusn. in Acta Horti Petrop. 15 (1898) 250; H. Smith in Hand.-Mazz. Symb. Sin. 7 (1936) 950; Marq. in Kew Bull. 3 (1937) 153; Hao in Bot. Jahrb. 68 (1938) 630; Walker in Contribs U.S. Nat. Herb. 28 (1941) 652; Fl. Sin. 62 (1988) 150.

—Ic.: Fl. Sin. 62, tab. 24, fig. 6–10.

Described from Qinghai. Type in St.-Petersburg (LE). Plate III. fig. 2.

In alpine meadows, among shrubs, coastal meadows and willow groves, 2200–3950 m.

IIIA. Qinghai: *Nanshan* (Jugum N. a Fl. Tetung, in pratis alpinis parce Aug. 15 [27] 1872, No. 406, Przew.—typus !; vicinity of Dan'ger town and Chorten-tan monastery, 2150–2450 m, Aug. 26, 1901—Lad.; "Kokonor: am Fuss des östlichen Nan-Schan, 3300 m, 1930"—Hao, l.c.; Mon'yuan', valley of Ganshiga river, left tributary of Peishikhe river in region of stud farm, 3350–3780 m, 1958—Dolgushin).

IIIB. Tibet: *Weitzan* (Yantszy-tszyan basin, valley of Go-chyu brook, 3950 m, in scrubs, Aug. 24; same site, along banks of Go-chyu brook in willow thickets and on mountain descents, Aug. 24–1900, Lad.).

General distribution: China (Nor.-West.: Gansu, South-West.: Sichuan).

68

## Section 6. Chondrophyllae Bge.

38. G. grandiflora Laxm. in Novi Comment. Ac. Sci. Petrop. 18 (1774) 526; Grossh. in Fl. SSSR, 18 (1952) 572; Grub. Konsp. fl. MNR [Conspectus of Flora of Mongolian People's Republic] (1955) 224; Fl. Kazakhst. 7 (1964) 105; Grub. Opred. rast. Mong. [Key to Plants of Mongolia] (1982) 201; Pakhom. in Opred. rast. Sr. Azii [Key to Plants of Mid. Asia] 8 (1986) 43.—*G. altaica* Pall. Fl. Ross. 2 (1788) 109; Ldb. Fl. Ross. 3, 1 (1847) 61; Kusn. in Acta Horti Petrop. 15, 3 (1904) 352; Sap. Mong. Altai (1911) 377; Kryl. Fl. Zap. Sib. 9 (1937) 2188.

—Ic.: Laxm. l.c. tab. 6, fig. 1; Pall. l.c. tab. 97, fig. 1; Fl. Kazakhst. 7, Plate 12, fig. 6.

Described from West. Siberia (Altay). Type in St.-Petersburg (LE). Plate II, fig. 5.

In alpine grasslands, rocky placers, moraines.

IA. Mongolia: *Mong. Alt.* (head of Karatyr river [Chern. Kobdo] Aug. 3, 1908—Sap.).

General distribution: Jung.-Tarb.; West. Sib. (Altay), East. Sib. (Sayan., Daur.), Nor. Mong. (Fore Hubs., Hent., Hang.).

39. G. aperta Maxim. in Bull. Ac. Sci. St.-Petersb. 27 (1881) 500; Forbes et Hemsl. Index Fl. Sin. 2 (1890) 123; Kusn. in Acta Horti Petrop. 15, 3 (1904) 378; Marq. in Kew Bull. 3 (1937) 173; Fl. Sin. 62 (1988) 214.—*G. maximowiczii* Kanitz in Pl. Exped. Szechenyi in Asia centr. coll. (1891) 39, tab. 3; Kusn. in Bull. Ac. Sci. St.-Petersb. 35 (1894) 352.

—Ic.: Kanitz, l.c. tab. 3 (sub nom. *G. maximowiczii* Kanitz); Fl. Sin. 62, tab. 34, fig. 57.

Described from Qinghai. Type in St.-Petersburg (LE). Plate IV, fig. 2.

In wet and marshy coastal meadows, alpine meadows, 2000-4100 m.

IIIA. Qinghai: *Nanshan* ("Altin-gomba circa Sining-fu, 2750 m, July 7, 1879, Szechenyi"—Kanitz, l.c.; Jugum a. Nanshan ad m. Donkyru extensum, ad fl. Rako-gol, 10,000–11,000 ft in palude ad ripam frequens et gregaria, July 10 [22] 1880, No. 522—Przew., typus !; Cha-dzha gorge, wet spots along Cha-dzha river, May 12, 1890—Gr.-Grzh.; nor. slope of South. Kukunor mountain range along Naion-Khutul'-gol river, 4100 m, along alpine meadows, Aug. 2, 1894—Rob.).

General distribution: endemic.

40. G. aquatica L. Sp. pl. (1753) 229; Grossh. in Fl. SSSR, 18 (1952) 579; Grub. Konsp. fl. MNR [Conspectus of Flora of Mongolian People's Republic] (1955) 223, p.p.; Fl. Kazakhst. 7 (1964) 107; Ohwi, Fl. Japan (1978) 1098; Grub. Opred. rast. Mong. [Key to Plants of Mongolia] (1982) 203, p. min. p.; Pakhom. in Opred. rast. Sr. Azii [Key to Plants of Mid. Asia] 8 (1986) 44; Fl. Sin. 62 (1988) 215. — *G. humilis* Stev. in Mem. Soc. natur. Moscou, 3 (1812) 258; Ldb. Fl. Ross. 3 (1847) 63; Boiss. Fl. or. 4 (1875) 72; Hemsl. in J. Linn. Soc. London (Bot.) 30 (1894) 117; Deasy, In Tibet a. Chin. Turk. (1901) 398; Kusn. in Acta Horti Petrop. 15, 3 (1904) 379, p.p.; Danguy in Bull. Mus. nat. hist. natur. 17, 3 (1911) 340; Kryl. Fl. Zap. Sib. 9 (1937) 2189.

Described form Siberia. Type in London (Linn).

In wet and marshy meadows, along banks of rivers, brooks and springs, from foothills to upper belt of mountains.

IIA. Junggar: *Tien Shan* (Khotun-Sumbul district, in Bain-bulak, 2560 m, 6393, Aug. 6, 1958 — Lee et Chu).

IIIA. Qinghai: *Nanshan* (on Kukunor lake, July 16, 1880 — Przew.; "Tatong, alt. 2900 m, route de Kantcheou a Sining, a travers le Nan-Chan, July 9, 1908, Vaillant" — Danguy, l.c.), *Amdo* (along Churmyn river, 2750–2900 m, May 16; along Baga-Gorgi river, 2750 m, May 23–1880, Przew.).

69     IIIB. Tibet: *Chang Tang* ("Tsaidam, close to water at 16,200 ft, 1891, Thorold" — Hemsl. l.c.; "33° lat., 82°53′19″, 15,000 ft, Aug. 30, 1896" — Deasy, l.c.), Weitzan (Yangtze basin, on Khichyu river, 4250 m, July 11, 1900 — Lad.).

General distribution: Fore Balkh., Jung. Alat., Nor. Tien Shan; Caucasus, West. Sib. (south, Altay), East. Sib. (Ang.-Sayan., Daur.), Nor. Mong. (Fore Hubs., Hent., Hang.).

41. G. aristata Maxim. in Bull. Ac. Sci. St.-Petersb. 26 (1880) 497; Forbes et Hemsl. Index Fl. Sin. 2 (1890) 124; Kanitz in Szechenyi, Wissensch. Ergebn. 2 (1898) 716; Kusn. in Acta Horti Petrop. 15, 3 (1904) 390; Danguy in Bull. Mus. nat. hist. natur. 17, 3 (1911) 340; Marq. in Kew Bull. 3 (1937) 171; Hao in Bot. Jahrb. 68 (1938) 628; Fl. Xizang, 3 (1986) 957; Fl. Sin. 62 (1988) 189.

— Ic: Fl. Xizang, 3, tab. 364, fig. 5–7; Fl. Sin. 62, tab. 30, fig. 8–10.

Described from Qinghai. Type in St. Petersburg (LE).

In floodplain and alpine meadows, grassy swamps, scrubs, forest meadows and glades, long floors of gorges, 2700–4500 m.

IIIA. Qinghai: *Nanshan* (Jugum S. a fl. Tetung, suma regione alpina et in vallibus montium, frequens et gregaria, July 1 [13] 1872, No. 224 — Przew., typus !; "in montibus circa monasterium Kumbum, 3900 m ad fin. July 1879, Szecheny" — Kanitz, l.c.; along Rako-gol river, 3050 m, along brooks, July 22, 1880 — Przew.; south slope of Xining alps, Chan-kho river, July 1, 1890 — Gr.-Grzh.; "Ou-po, alt. 3500 m, July 6, 1908, Vaillant" — Danguy, l.c.; "s.w. of Sining, Shangsin-chuang, 2700 m, F. Ridley" — Marq. l.c.; "auf den Abhängen des süd-kokonorischen Gebirgszuges, 3800 m; auf dem östlichen Nan-schan, 2800–3300 m, 1930" — Hao, l.c.), *Amdo* (Mudzhik mountains, alpine belt, 3200 m, June 27, 1880 — Przew.; Guidui-sha river, June 21, 1890 — Gr.-Grzh.; "Amne-Matchin, 4500 m bis 5000 m, 1930" — Hao, l.c.).

IIIB. Tibet: *Weitzean* ("Sosyan' " — Fl. Xizang, l.c.).

General distribution: China (Nor.-West.: Gansu, South-West.).

42. G. burkillii H. Smith in Hand.-Mazz. Symb. Sin. 7 (1936) 953; Hara, Chater et Williams, Enum. flow. pl. Nepal, 3 (1982) 92; Fl. Xizang, 3 (1986) 962; Fl. Siu. 62 (1988) 222. – *G. pseudo-humilis* Burk. in J. Asiat. Soc. Bengal, n.s. 2 (1906) 313, non Makino (1904).

—Ic.: Fl. Xizang, 3, tab. 365, fig. 19–22; Fl. Sin. 62, tab. 35, fig. 11–13.

Described from Himalayas (west.). Type in London (K).

On meadowy slopes, coastal meadows, gorges, 3600–4300 m.

IIIA. Qinghai (? "west. Qinghai" – Fl. Sin. l.c.).

IIIB. Tibet: *South.* (Ali: "Pulan', Chzhada" – Fl. Xizang, l.c.).

General distribution: Himalayas.

43. G. caeruleo-grisea T.N. Ho in Bull. Bot. Research, 4, 1 (1984) 77; Fl. Sin. 62 (1988) 157.

—Ic.: T.N. Ho, l.c. tab. 4, fig. 10–15.

Described from Qinghai. Type in Beijing (PE).

In alpine meadows and meadowy slopes, 3600–4250 m.

IIIA. Qinghai (? – Fl. Sin. l.c.).

IIIB. Tibet (? – Fl. Sin. l.c.).

General distribution: China (Nor.-West.: Gansu, south-west.).

44. G. clarkei Kusn. in Acta Horti Petrop. 15, 3 (1904) 419; Fl. Xizang, 3 (1986) 964; Fl. Sin. 62 (1988) 211. – *G. pygmaea* Clarke in Hook. f. Fl. Brit. India, 4 (1883) 111, non Rgl. – *G. saginoides* Burk. in J. Asiat. Soc. Bengal, n.s. 2 (1906) 318.

Described from Himalayas (west.). Type in London (K).

70    In wet and marshy meadows, meadowy slopes, around residences, 3700–4600 m.

IIIA. Qinghai (? – Fl. Sin. l.c.).

IIIB. Tibet: *South.* ("Lhasa, Chzhada" – Fl. Xizang, l.c.).

General distribution: Himalayas, Karakorum.

45. G. crassuloides Bureau et Franch. in J. de Bot. 5 (1891) 104; Kusn. in Acta Horti Petrop. 15, 3 (1904) 414; Marq. in Kew Bull. 3 (1937) 170, 188; Hara, Chater et Williams, Enum. flow. pl. Nepal, 3 (1982) 92; Fl. Xizang, 3 (1986) 952; Fl. Sin. 62 (1988) 200.

—Ic.: Fl. Xizang, 3, tab. 364, fig. 10–12; Fl. Sin. 62, tab. 32 fig. 5–7.

Described from South-West. China (Sichuan). Type in Paris (P).

On meadowy slopes, grassy marshes, among scrubs along banks of brooks and springs, 2700–4500 m.

IIIB. Tibet: *Weitzan* ("Tszichzhi" – Fl. Sin. l.c.).

General distribution: China (Nor.-West., South-West.), Himalayas.

46. G. crenulato-truncata (Marq.) T.N. Ho in Fl. Xizang, 3 (1986) 962; Fl. Sin. 62 (1988) 163. — *G. prostrata* var. *crenulato-truncata* Marq. in J. Linn. Soc. Bot. (London) 48 (1929) 205. — *G. prostrata* var. *bilobata* Marq. l.c. 205.

—Ic.: Fl. Sin. 62, tab. 27, fig. 16–20; Fl. Xizang, 3, tab. 365, fig. 7–9.

Described from Tibet. Type in London (K ?)

In alpine meadows, meadowy, rubbly-rocky and sand-covered slopes, along sides of mountain roads, floors of ravines and sand banks of lakes, 2700–5300 m.

IIIA. Qinghai (Fl. Sin. l.c.).

IIIB. Tibet: *Chang Tang* ("Gaitsze, Getszi, Zhitu, Shuankhu"), *Weitzan* ("Tszyali"), *South.* ("Lhasa, Pulan'")—Fl. Xizang, l.c.

General distribution: China (South-West.: nor. Sichuan, Kam).

47. G. grumii Kusn. in Acta Horti Petrop. 13, 4 (1893) 63 and 15, 3 (1904) 388; Gr.-Grzh. Zap. Kitai, 3 (1907) 493; Marq. in Kew Bull. 3 (1937) 173; Walker in Contribs U.S. Nat. Herb. 28 (1941) 651; Fl. Sin. 62 (1988) 190. — *G. maximowiczii* Kusn. in Bull. Ac. Sci. St.-Petersb. 34 (1891) 505, non Kanitz; id. in Acta Horti Petrop. 15, 3 (1904) 378. — *G. ivanoviczii* Marq. in Kew Bull. 3 (1937) 173.

—Ic.: Kusn. 1891, l.c. fig. 1–4; Gr.-Grzh. l.c.; Fl. Sin. 62, tab. 31, fig. 8–10.

Described from Qinghai. Type in St.-Petersburg (LE). Map 1.

In wet and marshy meadows, along banks and shoals of rivers forest borders, 3200–3600 m.

IIIA. Qinghai: *Nanshan* (South. Tetung mountain range, along valleys, April 28, 1873; along Tetung river, April–May 1873—Przew.; valle fl. Sining supra Ssiao-ssia, in palude, April 24, 1885—Pot., typus *G. maximowiczii* Kusn.; at Guman'sy temple, among pebbles on river bed, No. 93, May 5, 1890—Gr.-Grzh., typus !; "Hsimi-Jai [Ping Fan distr.], beautifully covering the edges of woods, 1923, R.C. Ching"—Walker, l.c.).

General distribution: endemic.

48. G. hyalina T.N. Ho in Bull. Bot. Research, 4, 1 (1984) 82; Fl. Xizang, 3 (1986) 967; Fl. Sin. 62 (1988) 159.

—Ic.: Fl. Xizang, 3, tab. 366, fig. 12–18.

Described from Tibet. Type in Beijing (PE).

In alpine meadows and along meadowy slopes, 4600–5300 m.

IIIA. Qinghai: ? (Fl. Sin. l.c.).

IIIB. Tibet: *Weitzan* ("Yuishu"), *South.* ("Chzhada")—Fl. Xizang, l.c.

General distribution: endemic.

49. G. karelinii Griseb. in DC. Prodr. 9 (1845) 106; Ldb. Fl. Ross. 3 (1847) 62; Grossh. in Fl. SSSR, 18 (1952) 578; Fl. Kirgiz. 8 (1959) 191; Ikonnik, Opred. rast. Pamira [Key to Plants of Pamir] (1963) 197; Fl. Kazakhst. 7 (1964) 106; Grub. Opred. rast. Mong. [Key to Plants of Mongolia] (1982) 202; Pakhom. in Opred rast. Sr. Azii [Key to Plants of

Mid. Asia] 8 (1986) 44; Fl. Sin. 62 (1988) 162. — *G. prostrata* β *karelinii* Kusn. in Acta Horti Petrop. 15, 3 (1904) 368. — *G. variabilis* Rupr. in Mem. Ac. Sci. St.-Petersb. ser. 7, 14, 4 (1869) 60. — *G. prostrata* auct. non Haenke;? Danguy in Bull. Mus. nat. hist. natur. 14 (1908) 131; ? Persson in Bot. notiser (1938) 297.

— Ic.: Fl. Kazakhst. 7, Plate 12, fig. 9.

Described from East. Kazakhstan. Type in Geneva (G).

In alpine meadows, syrts (watershed uplands), sasas (solonchaks on lateral seepages) and moraines, rock screes and rocks, along banks of brooks and on shoals, in spruce forests, 2000–5000 m.

IA. Mongolia: *Mong. Alt.* (nor.-west).

IIA. Junggar: *Tarb.* (north of Dachen town, Koktugai, 2250 m, No. 1566, Aug. 13, 1957 — Shen. same site, No. 2923, Aug. 13, 1957 — Kuan), *Tien Shan* (Sairam lake, July 12, 1877 — Fet.; same site, July 14 and Talkibash mountains, July 19 — 1877, A. Reg.; Urten-Muzart pass, Aug. 2, 1877; Piluchi, June 21, 1878 — Fet.; upper Tekes, June 26; Mukhurdai river, 3650–3950 m, July 17–1893, Rob.; Uch-Turfan, Karagaili gorge, in spruce grove, June 20, 1908 — Divn.; Am Wege von Fucan zur Bogdo-Ola, Aug. 2–3, 1908 — Merzb; on south-east. bank of Sairam lake, 2150 m, Aug. 18, 1953 — Mois.; in Danu region, No. 1460, July 17; nor. slopes in Danu-Daban region, 2200 m, No. 1946, July 18, 1957, Kuan; 8 km south of Nyutsyuan'tsza, 2000 m, No. 660, July 19, 1957 — Shen; Savan district, in Datszymyao region, No. 1712, July 22, 1957 — Kuan; Dzhagastai, 10 km south of pit toward Chzhaosu, 2400 m, in forest, No. 698, Aug. 8; Syata-Ven'tsyuan', 2200–2900 m, along bank of river, No. 1422, Aug. 13–1957, Shen; between Syata and Ven'tsyuan', 2000 m, No. 3410, Aug. 13; 10 km nor. of Chzhaosu, 2980 m, No. 3316, Aug. 15–1957, Kuan, Kulangou?, 2180 m, No. 945, Aug. 16; nor. bank of Sairam lake, 2800 m, No. 2121 and 2123, Aug. 27–1957, Shen; Ulyastai district, in Yakou town region, No. 3964, Aug. 30, 1957 — Kuan; north of Yakou town [Barchat], 2580 m, in forest, No. 1956, Aug. 31; 15 km south of Tyan'chi lake, 2550 m, No. 1955, Sept. 19–1957, Shen; Bortu to timber works in Khomote, 2300 m, No. 6998, Aug. 3; in Bain-bulak, 2560 m, No. 6393, Aug. 6 [together with *G. aquatica* L.] — 1958, Lee et Chu).

IIIA. Qinghai: *Nanshan* (mountains and Yamatyn-umru area, 3650–3950 m, July 23; Bukhyn-daban mountain range in upper Ara-gol river, 4250 m, Aug. 1–1894, Rob.).

IIIB. Tibet: *Chang Tang* (Kuen-Lun', Tokhtakhon mountains, nor. slope 3050–3650 m, alpine meadow and in spruce forest, July 21, 1889 — Rob.).

IIIC. Pamir (Ulug-tuz gorge in Charlym river basin, along bank of brook, June 26; same site, in estuary, July 1–1909, Divn.; Tashkurgan, in meadow, July 25, 1913 — Knorring; pass from Arpalyk river gorge in Kyzyl-Bazar, 3500 m, July 13; Mia river basin, 3300 m, July 22–1941; Goo-dzhiro river, 4500–5500 m, July 27, 1942 — Serp.).

General distribution: Jung-Tarb., Nor. and Cent. Tien Shan, East. Pam.; Mid. Asia (mount.).

50. G. leucomelaena Maxim. in Bull. Ac. Sci. St.-Petersb. 34 (1891) 305; id. Diagn. pl. nov. asiat. 8 (1893) 33; Kusn. in Acta Horti Petrop. 15, 3 (1904) 376; Rehder in J. Arn. Arb. 14 (1933) 28; Marq. in Kew Bull. 3 (1937) 173, 189; Kryl. Fl. Zap. Sib. 9 (1937) 2190; Persson in Bot. notiser (1938) 297; Grossh. in Fl. SSSR, 18 (1952) 578; Grub. Konsp. fl. MNR [Conspectus of Flora of Mongolian People's Republic] (1955) 224; Fl. Kirgiz. 8 (1959) 191; Ikonnik. Opred. Rast. Pamira [Key to Plants of Pamir] (1963) 199; Fl. Kazakhst. 7 (1964) 106; Grub. Opred. rast. Mong.

72  [Key to Plants of Mongolia] (1982) 202; Hara, Chater et Williams, Enum. flow. pl. Nepal, 3 (1982) 92; Pakhom. in Opred. rast. Sr. Azii [Key to Plants of Mid. Asia] 8 (1986) 45; Fl. Xizang, 3 (1986) Fl. Sin. 62 (1988) 212. — *G. aquatica* auct. non L.: Clarke in Hook. f. Fl. Brit. India, 4 (1883) 110, p.p. — *G. humilis* auct. non Stev.: Clarke, l.c. 111.

— Ic.: Maxim. 1891, l.c. fig. 5–10; Fl. Xizang, 3, tab. 365, fig. 10–12; Fl. Sin. 62, tab. 35, fig. 1–3.

Described from Qinghai. Type in St.-Petersburg (LE).

In coastal meadows, grassy swamps and sasas [solonchaks on lateral seepages], in solonetzic meadows, along riverine valleys, along banks of brooks, around springs, 1500–5000 m.

IA. Mongolia: *Mong. Alt.* (along Tsitsirin-gol river, July 10, 1877 — Pot.; bank of Bodunchi river, July 19, 1898 — Klem.; Tsagan-gol river near Muzdy-bulak estuary, June 28, 1899 — Lad.; valley of Suok river, June 18, 1908 — Sap.; Uenchiin-gol basin, Khargaityn-gol valley 5 km above estuary, Aug. 14, 1970 — Grub., Dariima, Muld.), *Cent. Khalkha* (vicinity of Kholt area, arid river bed and bank of Kholt brook, Aug. 1–3, 1926 — Gus.; sand along south. fringe of Tsagan-nur lake, June 25, 1940 — Yun.), *Depr. Lakes* (01'ge-nur, on marshy mounds, Aug. 29, 1879 — Pot.), *Val. Lakes* (Bayan-nur lake 10 km south-west of Mandal somon, Aug. 10, 1945 — Yun.), *Gobi-Alt.* (Dzun-Saikhan mountain range, Ëlo creek valley, Aug. 23, 1931 — Ik.-Gal.).

IB. Qaidam: *Mount.* (Kurlyk-nor lake, Khoitu-Taryane area, 2750 m, June 1; Ichegyn-gol river, 3050 m, June 24–1895, Rob.; between Kurlyk and Toso-nor lakes, 2150 m, June 26, 1901 — Lad.), *Plains* (vicinity of Barun-tszasak khyrma, 2630 m, in solonchak swamps, June 26, 1901 — Lad.).

IIA. Junggar: *Jung. Gobi* (Khoni-Usuni-gobi, Gun-Tamga area, solonchak meadow, Aug. 2, 1947 — Yun.).

IIIA. Qinghai: *Nanshan* (alpes Nan-shan, 8000–8500 ft, ad fontes frequens, June 28 [July 10] 1879, No. 272 — Przew, typus !; valley of Runvyr river, May 24, 1885 — Pot.; Humboldt mountain range, nor. slope, Blagodatnyi spring, 3350–3650 m, May 26; nor. slope of Ritter mountain range, upper Baga-Khaltyn-gol river, 4100 m, June 22; Ritter and Humboldt mountain ranges, 3650 m, July 2–1894, Rob.; Kuku-nor lake, Ui-yu area, lower belt, Aug. 13; Kuku-nor lake, Dzyao-dzyun-tai lake, Aug. 14—1908, Czet.), *Amdo* (Mudzhik mountains, 3500 m, June 27, 1880 — Przew.; "Ba valley, June 1926, Rock" — Rehder, l.c.).

IIIB. Tibet: *Chang Tang* (Keri mountain range, 3350–3650 m, rock fields, Aug. 5, 1885 — Przew.; Tokuz-Daban mountain range, nor. slope, 3350–3650 m, Muzlyk river, Aug. 1890 — Rob.; "Zhitu" — Fl. Xizang, l.c.), *Weitzan* (along Dyao-chyu river, July 11, 1884 — Przew.; Russkoe lake, 4100 m, along nor. bank of lake and along banks of Yellow river right at its head from lake, June 18; Dzhagyn-gol river, 4250 m, along banks, July 4; Yangtze river basin, along Khichyu river [vicinity of Nyamtso], 4250 m, July 11–1900; Burkhan-Budda mountain range, nor. slope, Khatu gorge, 3500 m, July 16, 1901 — Lad.), *South.* ("Lhasa, Nan'mulin, Pulan' " — Fl. Xizang, l.c.).

IIIC. Pamir (Ulug-tuz gorge in Charlym river basin, along slope, June 22; Pas-rabat river, Toili-bulun area, on brook, Aug, 2–1909, Divn.; "Pamir, Bostan-terek, about 3000 m, Aug. 13, 1934" — Persson, l.c.; Tas-pestlyk area, 4000–5000 m, July 25, 1942 — Serp.; Tash-kurgan, in valley, 3000 m, No. 291, June 13, 1959 — Lee et Chu).

General distribution: Jung. Alat.; Nor. and Cent. Tien Shan, East. Pam.; Mid. Asia (West. Tien Shan-Pam.-Alay), West. Sib. (Altay), East. Sib. (Ang.-Sayan., Daur.), Nor. Mong. (excepting Cis-Hing.), China (Nor.-West.: Gansu; South-West.: Sichuan), Himalayas.

51. G. ludlowii Marq. in Kew Bull. 3 (1937) 189; Hara, Chater et Williams, Enum. flow. pl. Nepal, 3 (1982) 92; Fl. Xizang, 3 (1986) 67; Fl. Sin. 62 (1988) 162.

—Ic.: Fl. Xizang, 3, tab. 365, fig. 13–15.

Described from South. Tibet. Type in London (BM).

In alpine meadows, shaded rocky slopes, mountain road sides, 3500–4700 m.

IIIB. Tibet: *South.* ("South. Tibet, Lhakang, 3900 m, shady rocky hillside, Sept. 1, 1933, No. 507, Ludlow et Sherriff–typus !" — Marq. l.c.; "Nan'mulin, Sage" — fl. Xizang, l.c.).

General distribution: China (South-West.: Kam), Himalayas (east.).

73   52. G. mailingensis T.N. Ho in Acta Biol. Plateau Sin. 3, 3 (1984) 30; Fl. Xizang, 3 (1986) 955; Fl. Sin. 62 (1988) 204.

—Ic.: T.N. Ho, l.c. tab. 5, fig. 2; Fl. Xizang, 3, tab. 352, fig. 6–10.

Described from South-West. China (Kam). Type in Beijing (PE).

In alpine meadows, borders of scrubs, 3900–4800 m.

IIIB. Tibet: *Weitzan* ("Bizhu" — Fl. Xizang, l.c.).
General distribution: China (South-West.: Kam).

53. G. micantiformis Burk. in J. Asiat. Soc. Bengal, n.s. 2, 7 (1906) 315; Marq. in Kew Bull. 3 (1937) 189; Hara, Fl. E. Himal. 2 (1971) 107; Fl. Xizang. 3 (1986) 956; Fl. Sin. 62 (1988) 188.

—Ic.: Fl. Xizang, 3, tab. 364, fig. 1–2.

Described from South. Tibet (Yatung). Type in London (K).

In alpine meadows and turf-covered slopes, 4300–4600 m.

IIIB. Tibet: *South.* ("Lankatsza" — Fl. Xizang, l.c.).
General distribution: Himalayas (east.).

54. G. nutans Bge. in Nouv. Mem. Soc. natur. Moscou, 1 (1829) 232; id. in Ldb. Fl. Alt. 1 (1829) 284; Kryl. Fl. Zap. Sib. 9 (1937) 2188; Grossh. in Fl. SSSR, 18 (1952) 577; Grub. Konsp. fl. MNR [Conspectus of Flora of Mongolian People's Republic] (1955) 224; Fl. Kazakhst. 7 (1964) 105; Grub. Opred. rast. Mong. [Key to Plants of Mongolia] (1982) 202; Fl. Sin. 62 (1988) 163. — *G. prostrata* auct. non Haenke: Kusn. in Acta Horti Petrop. 15, 1 (1904) 362;? Danguy in Bull. Mus. nat. hist. natur. 20 (1914) 75; Pakhom. in Opred. rast. Sr. Azii [Key to Plants of Mid. Asia] 8 (1986) 43.

—Ic.: Fl. SSSR, 18, Plate 30, fig. 2; Fl. Kazakhst. 7, Plate 12, fig. 7.

Described from Altay. Type in St.-Petersburg (LE).

In wet, marshy and coastal meadows, in *Cobresia* thickets, scrubs, larch groves, along banks of rivers and brooks, shaded rocks, mountain tundra, 1600–5000 m.

IA. Mongolia: *Khobd.* (Kharkhira mountain group, Kharkhira river, July 21, 1903 —
Gr.-Grzh.), *Mong. Alt.* (Ukha river, Aug. 6, 1899 — Lad.; Kak-kul' lake, between Tsagan-
gol and Kobdo, June 22, 1906; Dain-gol lake, south-west. bank, July 29, 1908 — Sap.;
Khara-Adzarga mountain range around Khairkhan-Duru, Aug. 25 and 26; Taishiri-ula,
nor. slope, Aug. 9–1930, Pob.; upper Indertiin-gol, summer camp site in Bulgan somon,
July 24, 1947 — Yun.; Dayan-nur lake, south. extremity, valley of brook 3 km above
settlement, July 8. 1971 — Grub., Dariima, Ulzij.; Khasagtu-Khairkhan mountain range,
Tsagan-Irmyk-ula, nor. slope in upper Khunkerin-ama, 2450–2500 m, Aug. 23, 1972 —
Grub., Ulzij. et al), *Cent. Khalkha* (Ubur-Dzhargalante river between Botogo and Agit
mountains, Aug. 30, 1925 — Krasch. et Zam.). *Val. Lakes* (Telin-gol bank, Aug. 15, 1926
— Gus.), *Gobi-Alt.* (Ikhe-Bogdo mountain range, south. slope, upper Khush' creek
valley, along floor of brook, Sept. 6, 1943 — Yun.; same site, Narim-Khurimt gorge,
around 3500 m, July 29; same site, in gorge mouth, 2450–2500 m, July 30–1948, Grub.).

IIA. Junggar: *Cis-Alt.* (Qinhe district, around Nun'tai settlement, 2600 m, No. 1085,
Aug. 4, 1956 — Ching), *Tien Shan* (east of Sarytyurk pass, on Koksai river, 2900 m, July
22; Bol. Yuldus, 2750 m, Aug. 5–1893, Rob.; 15 km south of Tyan'chi lake, 2550 m, No.
1955, Sept. 19, 1957 — Shen [together with *G. karelinii*]; Chzhaosu district, between
Ven'tsyuan' and Syata, 2100–2400 m, No. 4554, Aug. 15, 1957 — Kuan; Bol. Yuldus basin
30–35 km west — south-west of Bain-bulak settlement, border of floor, Aug. 10, 1958 —
Yun. et Yuan; in Aksu on Talakyus canal, 3100 m, No. 8460, Sept. 23, 1958 — Lee at Chu).

IIIA. Qinghai: *Nanshan* (at foot of Machan-ula, 3350–3650 m, July 23, 1879 — Przew.;
Humboldt mountain range, Ulan-bulak spring, 3650 m, July 5; Yamatyn-umru-ula
mountains, 3950 m, July 22–1894; nor. slope of Humboldt mountain range, Chon-sai
gorge, 3350 m, July 23, 1895 — Rob.; pass on Uvei-Lanzhou highway, 2800 m, Oct. 9, 1957
— Petr.).

IIIB. Tibet: Chang Tang (Keri mountain range, Ulug-achik area, 3400 m, Aug. 5,
1885 — Przew.).

IIIC. Pamir (near Yashilkan-Gunit settlement on Mia river, 4000 m, July 21, 1941;
Piyak-davan pass, 4500–5500 m, July 10; Tas-pestlyk area, 4000–5000 m, July 25–1942,
Serp.).

General distribution: Jung.-Tarb., Nor. and Cent. Tien Shan, East. Pamir ?; Mid. Asia
(mount.), West, Sib. (Altay), East. Sib., Far East (nor. and Kamchatka), Nor. Mong. (Fore
Hubs., Hent., Hang.).

## 55. G. prattii Kusn. in Acta Horti Petrop. 13 (1893) 63 and 15, 3 (1904) 387; Marq. in Kew Bull. 3 (1937) 174; Fl. Sin. 62 (1988) 208.

— Ic.: Fl. Sin. 62, tab. 34, fig. 1–4.

Described from South-West. China (Sichuan). Type in London (K).

In wet and marshy meadows, meadowy slopes, around residences,
3000–4000 m.

IIIA. Qinghai (? — Fl. Sin. l.c.).

General distribution: China (Nor.-West.: Gansu, Shenxi; South-West.: Sichuan,
Yunnan), Himalayas.

## 56. G. pseudo-aquatica Kusn. in Acta Horti Petrop. 13 (1893) 63 and 15, 3 (1904) 388; Strachey, Catal. (1906) 116; Gr.-Grzh. Zap. Kitai, 3 (1907) 493; Kryl. Fl. Zap. Sib. 9 (1937) 2191; Marq. in Kew Bull. 3 (1937) 173; Hao in Bot. Jahrb. 68 (1938) 629; Kitag. Lin. Fl. Mansh. (1939) 359; Grossh. in Fl. SSSR, 18 (1952) 580; Grub. Konsp. fl. MNR [Conspectus of flora of

84

Mongolian People's Republic] (1955) 225; Fl. Intramong. 5 (1980) 70;
Grub. Opred. rast. Mong. [Key to Plants of Mongolia] (1982) 202; Fl.
Xizang, 3 (1986) 965; Fl. Sin. 62 (1988) 221.—*G. aquatica* auct. non L.:
Clarke in Hook. f. Fl. Brit. India, 4 (1883) 110, p.p.; Franch. Pl. David. 1
(1884) 211; Forbes et Hemsl. Index Fl. Sin. 2 (1890) 123; Deasy, In Tibet
and Chin. Turk. (1901) 398; Hemsl. Fl. Tibet (1902) 190.

—Ic.: Fl. Intramong. 5, tab. 28, fig. 7–11; Grub. Opred. rast. Mong.
[Key to Plants of Mongolia] Plate 108, fig. 489; Fl. Sin. 62, tab. 35, fig.
9–10.

Described from East. Siberia. Type in St.-Petersburg (LE).

In wet and marshy meadows and hummocky swamps, coastal
meadows and meadowy slopes, 1500–5200 m.

IA. Mongolia: *Cent. Khalkha* (along east. bank of Ugei-nur lake, June 11, 1893 —
Klem.), *Gobi Alt.* (Dundu-Saikhan mountains, west. slope, July 1, 1909 — Czet.).

IIA. Junggar: *Tien Shan* (plain in Barkul' town, June 2–1879, Przew.).

IIIA. Qinghai: *Nanshan* (in Bamba village, April 2, 1884 — Przew.; Pabao-sy
monastery on Edzin-gol river, 2734 m, May 9; along Bardun river, 2700 m, May 12–1886,
Pot.; "Goman'sy monastery, 9400 ft, in pebbles along river bed, May 5, 1890" — Gr.-Grzh.
l.c.; nor. slope of Humboldt mountain range, vicinity of Kuku-usu area, 3650 m, June 6;
Yamatyn-umru mountains, 3950–4250 m, July 27–1894; South. Kukunor mountain range,
Dulan-gol, 3050 m, May 7, 1895 — Rob.).

IIIB. Tibet: *Chang Tang* (33°39′ lat., 82°45′57″ long., 16,800 ft. Aug. 20, 1896;
35°39′46″ lat., 82°0′6″ long., 16,000 ft. July 25, 1898" — Deasy, l.c.), *Weitzan* (Burkhan-
Budda mountain range, nor. slope, Nomokhun gorge, 3950 m, May 18; same site, 4250
m, May 20; Russkoe lake, along banks of lake and Yellow river, 4100 m, June 18; basin
of Yangtze river, Chzhabu-vrun area, 4250 m, along brook, July 10–1900; Russkoe lake,
4100 m, along mounds, June 2; Amnen-kor mountain range, south. slope, 4250 m, June
9; Burkhan-Budda mountain range, nor. slope, Khatu gorge, 4100 m, July 13–1901, Lad.;
"Sosyan'" — Fl. Xizang, l.c.), *South.* (Ali: "Gyanima, 13,000–15,000 ft, 1848" — Strachey,
l.c.; "Lhasa" — Fl. Xizang, l.c.).

General distribution: West. Sib. (south-east. Altay), East. Sib. (south.), Nor. Mong.
(excluding Cis-Hing.), China (Nor., Nor.-West., South-West., Cent.), Himalayas (west.,
Kashmir).

57. G. pudica Maxim. in Bull. Ac. Sci. St.-Petersb. 26 (1880) 497; Forbes
75 et Hemsl. Index Fl. Sin. 2 (1890) 133; Diels in Futterer, Durch Asien, Bot.
repr. (1903) 16; Marq. in Kew Bull. 3 (1937) 176; Hao in Bot. Jahrb. 68
(1938) 629; Fl. Sin. 62 (1988) 157.—*G. prostrata* ε *pudica* Kusn. in Acta
Horti Petrop. 15, 3 (1904) 370.

—Ic.: Fl. Sin. 62, tab. 27, fig. 1–5.

Described from Qinghai. Type in St.-Petersburg (LE).

In alpine meadows and meadowy slopes, coastal meadows, among
scrubs, 2200–5000 m.

IIIA. Qinghai: *Nanshan* (Regio alpina jugi N. a. fl. Tetung pratis alpinis frequens,
No. 400, Aug. 6 [18] 1872 — Przew., typus !; pratis alpinis jugi finitiam borealis frequens,

Aug. 20, 1880 — Przew.; South. Kukunor mountain range, nor. slope, 3600–3950 m, Godaban mountains, alpine meadow, Aug. 2, 1894 — Rob.; South. Kukunor mountain range, Tszaga-siten-këtël' pass, 3600 m, along mountain descents and alongside brooks, Aug. 17, 1901 — Lad.; Kuku-nor lake, Ui-yu area, in all belts of south-east. and east. slope, Aug. 13, 1908 — Czet.; 108 km west of Xining, 6 km west of Daudankhe settlement, 3400 m, mountain shrub steppe, Aug. 5, 1959 — Petr.; "Küke-nur, mitte Aug.; übergangspass in Süd-Küke-nur Gebirge, Aug. 27, Semenow-Gebirge, Sept. 7 — Futterer" — Diels, l.c.), *Amdo* ("Amne-Matchin, in den Tälern um 4500–5000 m, 1930" — Hao, l.c.).

General distribution: China (Nor.-West.: Gansu, Shanxi; South-West.: west. Sichuan).

58. G. riparia Kar. et Kir. in Bull. Soc. Natur. Moscou, 14 (1841) 706; Griseb. in DC. Prodr. 9 (1845) 106; Ldb. Fl. Ross. 3 (1847) 63; Kusn. in Acta Horti Petrop. 15, 3 (1904) 417; Hand.-Mazz. in Österreich. Bot. Zeitschr. 79 (1930) 37; Rehder in J. Arn. Arb. 14 (1933) 29; Marq. in Kew Bull. 3 (1937) 175; Kryl. Fl. Zap. Sib. 9 (1937) 2192; Grossh. in Fl. SSSR, 18 (1952) 581; Fl. Kirgiz. 8 (1959) 192; Fl. Kazakhst. 7 (1964) 107; Grub. Opred. rast. Mong. [Key to Plants of Mongolia] (1982) 202; Pakhom. in Opred. rast. Sr. Azii [Key to Plants of Mid. Asia] 8 (1986) 45; Fl. Sin. 62 (1988) 250.

— Ic.: Fl. Kazakhst. 7, Plate 12, fig. 8.

Described from East. Kazakhstan (Fore Balkhash). Type in Moscow (MW).

In coastal wet, solonetzic and swampy meadows, hummocky swamps, shoals, foothills to upper belt of mountains.

IA. Mongolia: *Mong. Alt.* (around Undur-Khairkhan mountains, Dundu-Tseren-gol river, along coastal pebble bed, Sept. 16, 1930 — Pob.).

IB. Kashgar: *Nor.* (Uch-Turfan, Kukurtuk gorge, along bank of spring, May 27, 1908 — Divn.).

IIIA. Junggar: *Jung. Gobi* (nor. foothills of Mechin-ula mountain range, Santakhu village, along solonetzic sandy bank of brook, May 11, 1877 — Pot.; Borotsonchzhi area, solonchaklike meadow, Sept. 13, 1948 — Grub.).

IIIA. Qinghai: *Amdo* ("Radja and Yellow river gorges, June 1926, Rock" — Rehder, l.c.).

IIIB. Tibet: *Chang Tang* ("Südufer des Mangza-ka 5370 m, 1906, Zugmeyer" — Hand.-Mazz. l.c.).

General distribution: Fore Balkh., Jung.-Tarb., Nor. and Cent. Tien Shan; Mid. Asia (Pam.-Alay), West. Sib., East. Sib. (Ang.-Sayan., Daur.), China (Nor.-West.).

59. G. simulatrix Marq. in Kew Bull. 3 (1937) 190; Fl. Xizang, 3 (1986) 952; Fl. Sin. 62 (1988) 202.

— Ic.: Fl. Xizang, 3, tab. 352, fig. 17-25.

Described from South. Tibet. Type in London (BM).

In alpine meadows and meadowy slopes, among shrubs, coastal meadows and shoals, 3000–4000 m.

IIIA. Qinghai (? — Fl. Sin. l.c.).

IIIB. Tibet: *South.* ("Abundant on grassy hill slopes, 3000 m, May 18, 1933 ?, No. 12, Ludlow et Sherriff, typus ! — Marq. l.c.; "Lhasa, Tszyantsze [Gyantze]" — Fl. Xizang, l.c.).

86

General distribution: endemic.

60. **G. spathulifolia** Maxim. ex Kusn. in Bull. Ac. Sci. St.-Petersb. 35 (1894) 351; Kusn. in Acta Horti Petrop. 15, 3 (1904) 386; Gr.-Grzh. Zap. Kitai, 3 (1907) 493; H. Smith in Hand.-Mazz Symb. Sin. 7 (1936) 959; Marq in Kew Bull. 3 (1937) 173; Fl. Sin. 62 (1988) 218. – *G. aperta* Maxim. Diagn. pl. nov. asiat. 8 (1983) 32, p.p.

– Ic.: Kusn. in Mel. Biol. 13, fig. 53–55; Fl. Sin. 62, tab. 35, fig. 4–8.

Described from Nor.-West. China (Gansu). Type in St.-Petersburg (LE).

In wet and coastal meadows, meadowy slopes, 2700–3800 m.

IIIA. Qinghai: *Nanshan* (Cha-dzhi gorge, meadowy sites along Cha-dzhi brook, 2750 m, May 12, 1890 – Gr.-Grzh.).

IIIB. Tibet: *Weitzan* (mountains along Bo-chyu river [tributary of Yangtze river], June 15; mountains along Bydzhun river, July 4–1884, Przew.).

General distribution: China (Nor.-West.: south. Gansu; South-West.: nor. Sichuan).

61. **G. squarrosa** Ldb. in Mem. Ac. Sci. St.-Petersb. 5 (1812) 520; ej. Fl. Ross. 3 (1847) 84; Trautv. in Acta Horti Petrop. 1 (1872) 185; Franch. Pl. David. 1 (1884) 211; Forbes et Hemsl. Index Fl. Sin. 2 (1980) 135, p. min. p.; Hemsl. Fl. Tibet (1902) 191; Kusn. in Acta Horti Petrop. 15, 3 (1904) 410; Diels in Futterer, Durch Asien, Bot. repr. (1903) 16; Danguy in Bull. Mus. nat. hist. natur. 20 (1914) 76; Marq. in Kew Bull. 3 (1937) 170; Kryl. Fl. Zap. Sib. 9 (1937) 2192; Kitag. Lin. Fl. Mansh. (1939) 359; Walker in Contribs U.S. Nat. Herb. 28 (1941) 652; Grossh. in Fl. SSSR, 18 (1952) 581, p. min. p; Grub. Konsp. fl. MNR [Conspectus of Flora of Mongolian People's Republic] (1955) 225; Fl. Kirgiz. 8 (1959) 192; Fl. Kazakhst. 7 (1964) 107; Ohwi, Fl. Japan (1978) 1098; Fl. Intramong. 5 (1980) 70; Grub. Opred. rast. Mong. {Key to Plants of Mongolia] (1982) 202; Pl. vasc. Helanshan (1986) 201; Pakhom. in Opred. rast. Sr. Azii [Key to Plants of Mid. Asia] 8 (1986) 45; Fl. Sin. 62 (1988) 197.

– Ic.: Fl. Intramong. 5, tab. 28, fig. 1–6; Fl. Sin. 62, tab. 32, fig. 1–4.

Described from East. Siberia (Trans-Baikal). Type in St.-Petersburg (LE).

On meadowy and steppe slopes, meadows, river terraces and rocky slopes, river banks and shoals, from plains to high mountains.

IA. Mongolia: *Cent. Khalkha* (along east. bank of Ugei-nur, July 11, 1893 – Klem.; Ubur-Dzhargalante river between sources and Agit mountain, Aug. 11; upper course of Ubur-Dzhargalante river between Botoga and Agit mountains, meadow along Ubur-Duryn river, Sept. 13–1925, Krasch. et Zam; vicinity of Ikhe-Tukhum-nor lake, Khairkhan valley, June 1926 – Zam.; south-east. foothills of Hangay, Kholt area, Aug. 11, 1926 – Gus.), *East. Mong.* (vicinity of Tsagan-Balgasu, May 1, 1831 – Ladyzh.; inter Kulussutaiewsk et Dolon-nor, No. 53, 1870 – Lomonossow; Muni-ula, declivitas borealis in fruticetis ad latera montium, June 25; ad. pedem Muni-ula australem, July–1871, Przew.; Kailar, steppe sablonneuse, alt. 750 m, June 22, 1896, Chaff." – Danguy, l.c.; Abder river, June 26; Ikhyr lake, July 6–1899, Pot. et Sold.; Yugodzyr, 15 km south-west of Erdene-Tsagan somon centre, alt. 1288 m, Aug. 12, 1970—Grub., Ulzij. et al; Shiliin-Bogdo-ula, nor. slope, July 8, 1971—Dashnyam, Karam.), *Alash. Gobi* (Alashan mountain range, desertified nor. foothills, July 21, 1873—Przew.; same site, Yamato gorge, west. slope, lower belt, May 5, 1908—Czet.; "Halahu-kou, on a moist valley bottom [on the north-west. side of Holanshan range], 1923, R.C. Ching"—Walker, l.c.).

IIA. Junggar: *Tien Shan* (bei Karabagh vor dem Eingang in das südliche Musart Tal auf alten Moränen, ende May 1903—Merzb.; 6 km south of Chzhaosu, 1760 m, No. 815, Aug. 11, 1957—Shen; Savan district, vicinity of Datszymyao, No. 1273, July 8, 1957—Kuan).

IIIA. Qinghai: *Nanshan* ("Am Küke-nur, No. 116, mitte Aug., Futterer"—Diels, l.c.; on way to Kukunor lake from Alashan, Shibanguku area, on bank of sandy river bed, Aug. 4, 1908—Czet.).

IIIB. Tibet: *Chang Tang* ("Tsaidam, banks of dry rivulet on hillside at 17,200 ft., 1891, Thorold"—Hemsl. l.c.), *Weitzan* (Burkhan-Budda mountain range, nor. slope, Khatu gorge, 3200 m, July 16, 1901—Lad.).

General distribution: Jung.-Tarb. (Tarb.), Nor. and Cent. Tien Shan; Mid. Asia (Badakhshan), West. Sib. (Altay), East. Sib. (south), Far East (south.), Nor. Mong., China (Nor., Nor.-East., Nor.-West.).

## 62. G. syringea T.N. Ho in Acta Biol. Plateau Sin. 3, 3 (1984) 22; Fl. Sin. 62 (1988) 192.

—Ic.: T.N. Ho, l.c. tab. 2, fig. 4.

Described from South-West. China (Sichuan). Type in Beijing (PE).

In alpine meadows and meadowy slopes, coastal meadows, 2200–3900 m.

IIIB. Tibet: *Weitzan* (Fl. Sin. l.c.).

General distribution: China (Nor.-West: south. Gansu; South-West: nor. Sichuan).

## 63. G. tatsienensis Franch. in Bull. Soc. Bot. France, 43 (1896) 489; Marq. in Kew Bull. 3 (1937) 177; Fl. Xizang, 3 (1986) 965; Fl. Sin. 62 (1988) 220.

Described from South-West. China (Sichuan). Type in Paris (P).

Along floors of gorges, riverine shoals, roadsides, 3300–4000 m.

IIIB. Tibet: *Weitzan* ("Sosyan'"—Fl. Xizang, l.c.).

General distribution: China (South-West.: west. Sichuan, Kam).

## 64. G. tetrasticha Marq. in Hook. Ic. pl. 34 (1937) tab. 3330, fig. 1–7; Fl. Xizang, 3 (1986) 968; Fl. Sin. 62 (1988) 160.

Described from south. Tibet. Type in London (K).

On meadowy slopes, 4200–5300 m.

IIIB. Tibet: *South.* ("Kuma, 4200 m, No. 115 C, 1935, Cutting et Vernay"—typus !—Marq. l.c.).

General distribution: endemic.

## 65. G. tricolor Diels et Gilg in Futterer, Durch Asien, Bot. repr. 3 (1903) 15; Fl. Sin. 62 (1988) 190.

—Ic.: Diels, l.c. tab. 1 C.

Described from Qinghai. Type in Berlin (B).

In moist and swampy meadows in floodplains of rivers and along banks of lakes, in forest meadows and borders, 2200–3200 m.

IIIA. Qinghai: *Nanshan* ("Küke-nur, mitte Aug. No. 90 a, 1898, Futterer, typus !— Diels, l.c.).

General distribution: China (Nor.-West.: west. Gansu).

## Subgenus 2. Gentianella (Moench) Kusn.
## Section 7. Crossopetalum Froel.
### (*Gentianopsis* Ma in Acta Phytotax. Sin. 1, 1 (1951) 7)

66. G. barbata Froel. Gentian. Diss. (1796) 114; Ldb. Fl. Ross. 3 (1847) 59; Danguy in Bull. Mus. nat. hist. natur. 14 (1908) 131 and 20 (1914) 75; Kryl. Fl. Zap. Sib. 9 (1937) 2178; Persson in Bot. notiser (1938) 297; Kitag. Lin. Fl. Mansh. (1939) 358; Grossh. in Fl. SSSR, 18 (1952) 595; Grub. Konsp. fl. MNR [Conspectus of Flora of Mongolian People's Republic] (1955) 223; Fl. Kirgiz. 8 (1959) 193; Fl. Kazakhst. 7 (1964) 110; Hilbig et Schamsran in Feddes Repert. 91 (1980) 39; Grub. Opred. rast. Mong. [Key to Plants of Mongolia] (1982) 201; Pakhom. in Opred. rast. Sr. Azii [Key to Plants of Mid. Asia] 8 (1986) 46.— *G. stricta* Klotzsch, Bot. Ergebn. Reise Prinz. Wald. Preuss. (1802) 90.— *G. vvedenskyi* Grossh. in Fl. URSS, 18 (1952) 751 and 597; Fl. Kirgiz. 8 (1959) 194; Ikonnik. Opred. rast. Pamira [Key to Plants of Pamir] (1963) 198.— *G. detonsa* auct. non Rottb.: Clarke in Hook. f. Fl. Brit. India, 4 (1883) 118, excl. var.; Franch. Pl. David. 1 (1884) 210; Forbes et Hemsl. Index Fl. Sin. 2 (1890) 127; Alcock, Rep. natur. hist. results Pamir bound. commiss. (1898) 24; Diels in Futterer, Durch Asien, 3 (1903) 16; Hao in Bot. Jahrb. 68 (1938) 629; Persson in Bot. notiser (1938) 297.— *Gentianopsis barbata* (Froel.) Ma in Acta phytotax. Sin. 1 (1951) 8; Fl. Intramong. 5 (1980) 78; Pl. vasc. Helanshan (1986) 201; Fl. Xizang, 3 (1986) 969; Fl. Sin. 62 (1988) 299.

—Ic.: Fl. Intramong. 5, tab. 33, fig. 1–5; Fl. Sin. 62, tab. 49, fig. 6–9.

Described from West. Siberia (Tom' river). Type in Berlin (B).

In moist coastal, often solonetzic meadows and pebble beds, larch forests and forest borders, willow and birch groves, alpine mountain meadows, 1000–4600 m.

IA. Mongolia: *Khobd.* (Ulan-daban pass on east. slopes, July 28, 1899—Lad.; south. slope of Tannu-01, Khorul'ma river, Aug. 1, 1903; Kharkhira mountain group, Kharkhira river, July 21, 1903—Gr.-Grzh.; Bukhu-Muren-gol floodplain 4–5 km north of same-named somon, July 31, 1945—Yun.; marshy zone between Bukhu-Muren-gol and Khub-Usu-gol rivers 7–8 km east-nor.-east of Bukhu-Muren somon, July 15, 1971—Grub., Ulzij., Dariima; south-west. spur of Turgen' mountain range, nor. slope of Yamat-ula mountain, July 18; Umne-Oger-ula 30 km nor.-east of Khobdo somon, 2550 m, July 30-1977, Karam. et al), *Mong. Alt.* (bank of Dzhirgalante river, running between Naryn and Shadzagai rivers, July 20, 1898—Klem.; Dain-gol lake, steppe slopes, Aug. 5, 1906—Sap.; Khasagtu-Khairkhan mountains, Dundu-Tseren-gol river near Undur-Khair-khan mountains, Sept. 16; same site, Sept. 17–1930, Pob.; lower course of Turgen-gol river, left bank tributary of Bulgan river, July 21, 1947 – Yun.), *Cent. Khalkha* (vicinity of Ikhe-Tukhum-nor lake, June 1926 – Zam.; valley of mid-course of Kerulen river, July 1899 – Pal.), *East. Mong.* (Kulun-Buir-nor plain, Bain-gol river, Aug. 7, 1899 – Pot. et Sold.; Khuntu somon 5 km west of Toge-gol river, Aug. 7, 1949 – Yun.; Kerulen river floodplain south of Enger-Shanda, Aug. 18, 1957 – Dashnyam; Shilin-Khoto town, true steppe, 1960 – Ivan.), *Depr. Lakes* (south. bank of Khara-Usu lake, Aug. 18; Kobdo river below Torkhul mountains, Aug. 25; vicinity of Ulangom, Sept. 6–1879, Pot.; lower Buyantu-gol, 1941 – Kondratenko; "Chovd-gol Aue an den Ongozny-Ulaan uul, Weiderasen" – Hilbig et Schamsran, l.c.), *Val. Lakes* (Gobi bank of Ologoi river, July 23, 1924 – Gorbunova; vicinity of Kholt area near Sondzhi, July 7, 1926 – Gus.), *Gobi-Alt.* (Ikhe-Bogdo mountains, nor. slope, upper belt, Aug. 23, 1926 – Tug.; Dzun-Saikhan mountains, on brook in Ëlo creek valley, Aug. 23, 1931 – Ik.-Gal.; Dundu- and Dzun-Saikhan mountains, July–Aug. 1933 – Simukova; Ikhe-Bogdo mountains, nor. slope, Ulyaste creek valley, July 1, 1972 – Banzragch), *Alash. Gobi* (Alashan mountain range, central portion, nor. slope of valley in midbelt, July 17, 1873 – Przew.), *Ordos* (between Shar-Burdyu area and Bain-nor lake, in dune sand among willow groves, Sept. 1, 1884 – Pot.).

IC. Qaidam: *Mount.* (vicinity of Dulan-khit temple along nor. bank of Tsagan-nor lake, 3100 m, Aug. 12, 1901 – Lad.).

IIA. Junggar: *Cis-Alt.* (Altay, 1900 m, entre 1'Ouchte et 1'Irtich, Aug. 20, 1895, Chaff. " – Danguy, l.c.), *Jung. Alat.* (mountains in Toli district, No. 2590, Aug. 7; vicinity of Ven'tsyuan' town, No. 3446, Aug. 13; 20 km east of Ven'tsyuan' town, No. 4670, Aug. 27–1957, Kuan; near Ven'tsyuan' town, No. 1810, Aug. 22, 1957 – Shen), *Tien Shan* (several finds), *Jung. Gobi* (Khoni-Usuni-Gobi, Gun-Tamga area, saline meadow, Aug. 2, 1947 – Yun.; Baitag-Bogdo mountain range, Aug. 1, 1988 – Gub., Kam. et al).

IIIA. Qinghai: *Nanshan* (nor. vicinity of Pinfan town, July 13, 1875 – Pias.; alpine belt, 2450–2600 m, bank of Dan-khe river, July 7; Nanshan alpine belt [foot of Machan-ul], 3350–3650 m, July 25–1879, Przew.; nor. slope of Humboldt mountain range, Tsagan-bulak area, 3650 m, July 5; Sharagol'dzhin village, Kuntun area, 3650 m, July 12; along Sukhai-khe river, 3650–3950 m, July 30–1894; nor. slope of Humboldt mountain range, Chon-sai gorge near pass, 3650 m, July 22, 1895 – Rob.; South. Kukunor mountain range, nor. slope under Tszagasyten-ketel' pass, 3450–3600 m, Aug. 17, 1901 – Lad.; on way to Kukunor lake, Shibanguku area, Aug. 4; Kukunor lake, Ui-yu area, Sept. 12–1908; along bed of Pinfan-khe river, July 21, 1909 – Czet.; "bei Kloster Kumbum, östlich Sining-fu" – Diels, l.c.), *Amdo* (along Mudzhik-khe river, 2750 m, June 29, 1880 – Przew.; "Kokonor: zwischen Schengenzogia und Gangia-zo, 3440 m" – Hao, l.c.).

IIIB. Tibet: *Chang Tang* (Tokhtakhon mountains, nor. slope, 3050–3650 m, July 21, 1889 – Rob.), *Weitzan* (Burkhan-Budda mountain range along Nomogun-gol river, Aug. 15, 1884 – Przew.; same site, nor. slope, Khatu gorge, 3200 m, July 8; same site, 3950 m, July 12, 1901 – Lad.; Yantszy-tszyan river, right bank, Nru-chyu area, 3550 m, July 25, 1900 – Lad.; "Sosyan'" – Fl. Xizang, l.c.), *South.* ("Chzhada, Pulan'" – Fl. Xizang, l.c.).

IIIC. Pamir (at confluence of Ulug-tuz river with Charlym, July 1; Toili-bulun area on Pasrabat river, Aug. 2–1909, Divn.; Mia river basin, 3300 m, July 22; Pasbarat area, 2500–3000 m, Aug.-Sept.–1941; in midcourse of Kaplyk river, 2500–3200 m, July 12; Tyna river estuary at its confluence with Issyk-su river, 3200 m, July 19–1942, Serp.; "in open Pamir, 13,000–14,000 ft" – Alcock, l.c.; "Muztag-ata, vallee de Tor-bachi, alt. 3350 m, July 31, 1906, Lacoste" – Danguy, 1908, l.c.; "Bostan-terek, July 1921; Jerzil, 3200 m, July 14, 1930" – Persson, l.c.).

General distribution: Jung.-Tarb., Nor. and Cent. Tien Shan, East. Pam.; Arct. (Asian), Europe (Volga-Kama. and Trans-Volga), Mid. Asia (West. Tien Shan, Pam.-Alay), West. Sib., East. Sib., Far East (Okhotsk, Zee-Bur.), Nor. Mong., China (Altay, Dunbei, Nor., Nor.-West., Cent., South-West.), Himalayas.

67. G. paludosa Munro ex Hook. f. in Hook. Ic. Pl. 9 (1852) vol. 857.
— *G. detonsa* var. *paludosa* Hook. f. 1.c. — *G. detonsa* var. *stracheyi* Clarke
in Hook. f. Fl. Brit. India, 4 (1883) 118; Strachey, Catal. (1906) 116; Hao
in Bot. Jalub. 68 (1938) 629. *C. detonsa* var. *ovato-deltoidea* Burkill in J.
Asiat. Soc. Bengal, nov. ser. 2 (1906) 319; Hao in Bot. Jahrb. 68 (1938)
629. — *Gentianella paludosa* (Munro) H. Smith in Hand.-Mazz. Symb. Sin.
7 (1936) 980; Hara, Chater et Williams, Enum. flow. pl. Nepal, 3 (1982)
94. — *Gentianopsis paludosa* (Hook. f.) Ma in Acta phytotax. Sin. 1 (1951)
31; Fl. Xizang, 3 (1986) 969; Fl. Sin. 62 (1988) 294.

—Ic.: Hook. Ic. Pl. 9, tab. 857; Fl. Sin. 62, tab. 49, fig. 1–3.

Described from Tibet. Type in London (K).

In meadows, along banks of brooks, pebble beds, among scrubs,
irrigated fields, 1200–4300 m.

IIIA. Qinghai: *Nanshan* (vicinity of Choibsen monastery, July 17; South. Tetung
mountain range, south. slope, July 28–1872; on Rako-gol river, 3050 m, July 22, 1880—
Przew.; "am Ufer des Sees Kokonor, um 3300 m"—Hao, l.c.; 6 km west of Daudaikhe
settlement between Xining and Gunkho, 3400 m, Aug. 5, 1959—Petr.), *Amdo* (Guidui-
sha, May 7, 1885—Pot.; "auf dem Gebirge Jahe-mari, 3400–4000 m"—Hao, l.c.).

IIIB. Tibet: *Weitzan* (right tributary of By-chyu river [Dyao-chyu], 4200 m, July 13,
1884—Przew.; Yantszy-tszyan river basin, upper Khi-chyu river, Chzhabu-vrun area,
4250 m, July 10, 1900—Lad.; Burkhan-Budda mountain range, nor. slope, Khatu gorge,
3200 m, June 24; same site, 4250 m, July 12; same site, 4100 m, July 13; same site, Ikhe-
gol gorge, 3350–3650 m, July 23–1901, Lad.; "Sosyan'"—Fl. Xizang, l.c.), *South.*
("Chzhada, Nan'mulin, Mochzhugunka"—Fl. Xizang, l.c.).

General distribution: China (Nor.: Hebei, Nor.-West., South-West.), Himalayas.

## Section 8. Comastoma Wettst.

68. G. falcata Turcz. Cat. baical, No. 783 (1837) nomen; in Bull. Soc.
Natur. Moscou, 15 (1842) 404, descr.; Ldb. Fl. Ross. 3 (1847) 55; Hemsl.
in J. Linn. Soc. (London) Bot. 30 (1894) 117; Danguy in Bull. Mus. nat.
hist. natur. 14 (1908) 131; Kryl. Fl. Zap. Sib. 9 (1937) 2174; Hao in Bot.
Jahrb. 68 (1938) 629; Persson in Bot. notiser (1938) 297; Grossh. in Fl.
SSSR, 18 (1952) 617; Grub. Konsp. fl. MNR [Conspectus of Flora of
Mongolian People's Republic] (1955) 223; Ikonnik. in Dokl. AN Tadzh.
SSR, 20 (1957) 56; Fl. Kirgiz. 8 (1959) 195; Ikonnik. Opred. rast. Pamira
[Key to Plants of Pamir] (1963) 198; Fl. Kazakhst. 7 (1964) 113; Hilbig et
Schamsran in Feddes Repert. 91 (1980) 39; Grub. Opred. rast. Mong. [Key
to Plants of Mongolia] (1982) 202; Pakhom. in Opred. rast. Sr. Azii [Key
to Plants of Mid. Asia] 8 (1986) 49. — *G. tenella* var. *falcata* (Turcz.) Griseb.
Gen. et sp. Gentian. (1838) 249; Clarke in Hook. f. Fl. Brit. India, 4 (1883)
110. — *G. hedinii* Murbeck in Oesterr. Bot. Zeitschr. 49, 7 (1899) 241;
Hedin, S. Tibet, 6, 3 (1922) 48. — *G. cordisepala* Murbeck, l.c. 243; Hedin,
l.c. 48; Diels in Futterer, Durch Asien, 3 (1903) 17. — *G. trailliana* auct. non

Forrest: Hao in Bot. Jahrb. 68 (1938) 630. — *Comastoma falcatum* (Turcz.)
Toyokuni in Bot. Mag. Tokyo, 74 (1961) 198; Fl. Intramong. 5 (1980) 83;
Fl. Xizang, 3 (1986) 975; Pl. vasc. Helanshan (1986) 203; Fl. Sin. 62 (1988)
306.

— Ic.: Fl. SSSR, 18, Plate 32, fig. 2; Fl. Kazakhst. 7, Plate 13, fig. 4; Fl.
Intramong. 5, tab. 34, fig. 1–4; Fl. Sin. 62, tab. 50, fig. 4–6.

Described from East. Kazakhst. (Jung. Ala Tau). Type in Moscow
(MW). Isotype in St.-Petersburg (LE).

In meadows and grasslands, along banks of brooks and rivers, in
*Cobresia* groves, swamps, spruce groves, moraines of glaciers, moist
rubbly and rocky slopes in alpine belt, 2100–5300 m.

IA. Mongolia: *Khobd.* ("Charchira: Suslan-Chamar, hochalpine Schotterhalde, 2600
m " — Hilbig et Schamsran, l.c.), *Mong. Alt.* (slopes of Shadzagain-Suburga pass, July 22,
1898 — Klem.; valley of Bulgan-gol above somon, July-Aug. 1947 — M. Tarasova; Adzhi-
Bogdo mountain range, Burgasyn-daba pass between Indertin-gol and Dzuslangin-gol,
Aug. 6, 1947 — Yun.; Adzhi-Bogdo mountain range, 3300 m, Aug. 23, 1973 — Isach. et
Rachk.), *Gobi-Alt.* (Ikhe-Bogdo mountain range, Bityuten-ama, Aug. 12, 1927 — M.
Simukova; same site, Narin-khurimt-ama, about 3500 m, July 29; same site, plateaulike
crest of mountain range, about 3700 m, July 29–1948, Grub.; same site, south.
macroslope, 3500 m, in gorge, Aug. 4, 1973 — Isach. et Rachk.), *Alash. Gobi* (Alashan
mountains, end of Sept. 1871 — Przew.).

IC. Qaidam: *mount.* (Ritter mountain range, nor. slope, Baga-Khaltyn-gol valley,
3950 m, July 22, 1894 — Rob.; Dulan-khit temple, 3350 m, in spruce and juniper forests,
Aug. 8, 1901 — Lad.).

IIA. Junggar: *Cis-Alt.* (pass from Mal. Kairta to Mal. Ku-Irtys, July 17, 1908 — Sap.),
*Jung. Alat.* (Ventsyuan' district, No. 4649, Aug. 25, 1957 — Kuan), *Tien Shan* (several
finds).

IIIA. Qinghai: *Nanshan* (on Gadzhur mountain, North. Tetung mountain range,
alpine meadow, Aug. 16, 1872; Nanshan alps, 2450 m, along bank of Dan-khe river, July
5; same site, along Kuku-usu river, July 12–1879, Przew.; Cherik pass, Aug. 8, 1890 — Gr.-
Grzh.; Humboldt mountain range, nor. slope, Ulan-bulak area, 3650 m, July 5 and 7;
Yamatyn-umru mountains, 3950 m, July 15; same site, 3650 m, July 23–1894; Humboldt
mountain range, nor. slope, Chan-sai gorge, 2750–3050 m, July 23, 1895 — Rob.; Kuku-
Nor lake, south. bank of Nara-Saren-Khutul' pass, 3200 m, near Tsagan-Yan'pin town
ruins, Aug. 21, 1901 — Lad.; Sangin area [south of Dangertin], Aug. 12, 1908 — Czet.; "am
sudlichen Ufer des Kuke-nur" — Diels, l.c.; "Kokonor, auf dem Selgen, 4800 m; am Fuss
des ostlichen Nan-Schan, 3300 m — 1930" — Hao, l.c.; pass on Uvei-Lanzhou highway,
2800 m, Oct. 9, 1957 — Petr.).

IIIB. Tibet: *Chang Tang* (Keri mountain range [Syren-bulun brook], 3350–3650 m,
Aug. 3; same site [Ulug-achik area], 3350–3650 m, Aug. 5–1885, Przew.; Tokhtakhon
mountains, nor. slope, 3050–3650 m, July 21, 1889 — Rob.; "Tsaidam, marsh at 15,000 ft,
Thorold, 1891" — Hemsl. l.c.; "Kwenlun, Sarik-kol, 3649 m, Aug. 5, 1896 — Hedin, l.c.;
"Zhitu, Shuankhu" — Fl. Xizang, l.c.), *Weitzan* (valley of Alak Nor gol river, in gorge,
Aug. 12; nor. slope of Burkhun-Budda mountain range, alpine meadow, Aug. 13–1884,
Przew.; same site, Nomokhun gorge, 4250 m, May 25, 1900 — Lad.; Russkoe lake, bank,
4100 m, June 20, 1900; Burkhan-Budda mountain range, nor. and south. slopes, 4100–
4250 m, July 12; same site, Khatu gorge, 3950 m, July 12; same site, 3200 m, July 24–1901,
Lad.; "Bizhu" — Fl. Xizang, l.c.), *South.* ("Chzhunba" — Fl. Xizang, l.c.).

IIIC. Pamir (Hattammig davan, 5400–5450 m, Aug. 19, 1892–D. Rhins; "Muztag-ata, vallee de Tor-Bachi, 3800 m, July 31, 1906, Lacoste"–Danguy, 1908, l.c.; valley of Chicheklik, July 28, 1909–Divn.; "Jerzil, 3000 m, July 10; 3300 m, July 21–1930; Jerzil Jaleu, 3800 m, July 31, 1931; Bostan-terek, Aug. 5 1934" –Persson, l.c.; Tiyak-davan pass, 4500–5500 m, July 10; in waterdivide of Kaplyk river and brook east of it, 5500–6000 m, July 13; Tas-pestlyk area, 4000–5000 m, July 25; gorge of Shorluk river, 4000–5500 m, July 28; Tegeboin pass, 4200–4300 m, Aug.-Sept.–1942, Serp.; "Kungur mountain range, nor.-west. slope, on glacier moraine, 4800 m, Aug. 13–17, 1956–Dmitriev et al" –Ikonnik. 1957, l.c.).

General distribution: Jung.-Alat., Nor. and Cent. Tien Shan, East. Pam.; Mid. Asia (Pam.-Alay), West. Sib. (Altay), East. Sib. (Sayans, Daur.), Nor. Mong. (Fore Hubs.), China (Nor.: Hebei, Nor.-West., South-West.: nor.-west. Sichuan).

69. **G. pedunculata** Royle ex G. Don, Gen. Syst. Dichlam. Pl. 4 (1837) 182. – *Gentianella pedunculata* (Royle) H. Smith in Grana Palyn. 7 (1967) 107, 144; Hara, Chater et Williams, Enum. flow. pl. Nepal, 3 (1982) 94. – *Comastoma pedunculatum* (Royle) Holub in Folia Geobot. et Phytotax. Praha, 3 (1968) 218; Fl. Xizang, 3 (1986) 975; Fl. Sin. 62 (1988) 310. – *Gentiana tenella* auct. non Rottb.: Clarke in Hook. f. Fl. Brit. India, 4 (1883) 109; Hao in Bot. Jahrb. 68 (1938) 630. – *G. borealis* auct. non Bge.: Clarke, l.c.

Described from Kashmir. Type in Liverpool (LIV).

In alpine meadows, along banks of brooks, river shoals, 3200–4800 m.

IIIA. Qinghai: *Amdo* ("Kokonor: Jahe-mari auf den Abhängen, 1930"–Hao, l.c.).

IIIB. Tibet: *Chang Tang* ("Zhitu"–Fl. Xizang, l.c.), *South.* ("Pulan"–Fl. Xizang, l.c.).

General distribution: China (Nor.-West.: Gansu, South-West.: Sichuan, Yunnan), Himalayas.

70. **G. polyclada** Diels et Gilg in Futterer, Durch Asien, 3 (1903) 16; Diels in Filchner, Wissensch. Ergebn. 10, 2 (1908) 262. – *G. limprichtii* Grüning in Feddes Repert. 12 (1913) 308. – *Comastoma limprichtii* (Grüning) Toyokuni in Bot. Mag. Tokyo, 74 (1961) 198; Fl. Intramong. 5 (1980) 83; Pl. vasc. Helanshan (1986) 203. – *C. polycladum* (Diels et Gilg) T.N. Ho in Fl. Sin. 62 (1988) 309.

–Ic.: Diels et Gilg, l.c. tab. 3; Fl. Intramong. 5, tab. 35; Fl. Sin. 62, tab. 50, fig. 11–13.

Described from Qinghai. Type in Berlin (B)?

On meadowy slopes, along banks of brooks and river shoals, marshy meadows along lake basins, 2100–4500 m.

IA. Mongolia: *Alash. Gobi* (Alashan mountain range–"Helanshan"–Pl. vasc. Helanshan, l.c., Fl. Intramong. l.c.).

IIIA. Qinghai: *Nanshan* ("2 km östlich vom Küke-nur; am Nordfuss der Korallenkalkberge", Futterer–Diels et Gilg, l.c., typus !; Kuku-Nor lake, Sangyn area, along southern slopes, Aug. 12, 1908–Czet.).

General distribution: China (Nor.-West.: Gansu, Shanxi).

71. G. pulmonaria Turcz. in Bull. Soc. Natur. Moscou, 22 (1849) 317; in Flora (1834) 1 Beibl. 19, nomen; Ldb. Fl. Ross. 3 (1847) 55; Hao in Bot. Jahrb. 68 (1938) 689; Grossh. in Fl. SSSR, 18 (1952) 614; Grub. Konsp. fl. MNR [Conspectus of Flora of Mongolian People's Republic] (1955) 225; id. Opred. rast. Mong. [Key to Plants of Mongolia] (1982) 202. — *G. arrecta* Franch. ex Hemsl. in J. Linn. Soc. (London) Bot. 26 (1890) 124. — *G. holdereriana* Diels et Gilg in Futterer, Durch. Asien, Bot. repr. 3 (1903) 17. — *Comastoma pulmonarium* (Turcz.) Toyokuni in Bot. Mag. Tokyo, 74 (1961) 198; Fl. Xizang, 3 (1986) 975; Fl. Sin. 62 (1988) 307.

— Ic.: Grub. Opred. rast. Mong. [Key to Plants of Mongolia] Plate 108, fig. 487; Fl. Sin. 62, tab. 50, fig. 7–10.

Described from East. Siberia (Trans-Baikal). Type in St.-Petersburg (LE).

In meadows, along banks of brooks and rivulets, pebble beds, wet rubbly slopes, larch and willow forests in upper belt of mountains, 2000–4800 m.

IA. Mongolia: *Khobd.* (valley of Turgen river, 2100 m, larch grove, July 7, 1973 — Banzragch, Karam. et al).

IIIA. Qinghai: *Nanshan* (South. Tetung mountain range, in meadows, July 23, 1872; along Yusun-Khatyma river, 2750–3350 m, July 23; nor. slope of South. Tetung mountain range [Sodi-suruksum mountains], alpine meadow, Aug. 2–1880, Przew.; "östlich vom Küke-nur" — Diels, l.c.), *Amdo* ("Kokonor, auf dem Plateau Dahoba, um 4000 m, 1930" — Hao, l.c.).

IIIB. Tibet: *Weitzan* (Yantszy-tszyan river basin, vicinity of Kabchzhi-Kamba village on Khi-chyu river, about 3700 m, on pebble banks of rivulet, July 21, 1900 — Lad.; "Sosyan" — Fl. Xizang, l.c.), *South.* ("Nan'mulin" — Fl. Xizang, l.c.).

General distribution: East. Sib. (Sayans, Daur.), Nor. Mong. (Fore Hubs., Hent., Hang.), China (Nor.-West., South-West.).

72. G. tenella Rottb. in Acta Hafn. 10 (1770) 436; Griseb. in DC. Prodr. 9 (1845) 98; Henderson et Hume, Lahore to Jarkand (1873) 328, cum auct. Fries; Deasy, Tibet and Chin. Turk. (1901) 398, cum auct. Fries; Hemsl. Fl. Tibet (1902) 191; Danguy in Bull. Mus. nat. hist. natur. 20 (1914) 76, cum auct. Fries; Kryl. Fl. Zap. Sib. 9 (1937) 2174; Grossh. in Fl. SSSR, 18 (1952) 618; Grub. Konsp. fl. MNR [Conspectus of Flora of Mongolian People's Republic] (1955) 225; Fl. Kazakhst. 7 (1964) 113; Hilbig et Schamsran in Feddes Repert. 91 (1980) 39; Grub. Opred. rast. Mong. [Key to Plants of Mongolia] (1982) 202; Pakhom. in Opred. rast. Sr. Azii [Key to Plants of Mid. Asia] 8 (1986) 49; Gubanov, Kamelin et al. in Byull. Mosk. obshch. ispyt. prir., otd. biol. 95 (1990) 122. — *G. malyschevii* (V. Zuev) Gubanov et R. Kam. in Byull. Mosk. obshch. ispyt. prir., otd. biol. 92 (1987) 122. — *Gentianella malyschevii* V. Zuev in Bot. zhurn. 70, 10 (1985) 1401. — *Comastoma tenellum* (Rottb.) Toyokuni in Bot. Mag. Tokyo, 74 (1961) 198; Fl. Xizang, 3 (1986) 976; Fl. Sin. 62 (1988) 309.

— Ic.: Acta Hafn. 10, tab. 2, fig. 6; Fl. Kazakhst. 7, Plate 13, fig. 5.

94

Described from West. Europe. Type in Copenhagen (C).

In grasslands and sasas (solonchaks), larch forests and forest glades, coastal meadows and pebble beds, moss-lichen tundra, rock fissures, 2000–5000 m.

IA. Mongolia: *Khobd.* (Kharkhira mountain range along Tarbagatai river, July 26, 1879–Pot.; south. slope of Sailyugem mountain range 70 km nor.-west of Ulan-Khus settlement, Khara-Dzhamash-gol basin [left tributary of Oiguriin-gol tributary], July 10, 1988–Gub., Dariima et al), *Mong. Alt.* (Khan-Taishiri mountain range, nor. slope, Aug. 9, 1930–Pob.; upper Indertiin-gol near Bulgan somon summer camp, July 24; Adzhi-Bogdo mountain range, upper Ubur-Dzuslan-gol creek valley, alpine belt, Aug. 7–1947, Yun.; Dayan-nur lake, nor. slope of Yamatyn-Ula mountains, 2350 m, July 10, 1971–Grub., Dariima et al; bank of Chigirtei-gol river, Aug. 6, 1979–Gub., Dariima, Kamelin; Sanginin-gol basin 25 km south-east of Dayan-nur post, 1800–2100 m, July 16–17; Ëlt-gol basin 30 km south of Altay settlement, July 22; Dzhangyz-Agach river basin, valley under Akhuntyiin-daba pass, July 26–1988, Kamelin, Ganbold et al.; "Erden-Buren Sum: Aue des Uljaschtain-gol nahe Dewseg"–Hilbig et Schamsran, l.c.), *Gobi-Alt.* (Ikhe-Bogdo mountain range, nor. slope, Ulyaste creek valley, 2800 m, July 2, 1972–Banzragch, Karam et al).

83 IIA. Junggar: *Cis-Alt.* ("montagues entre l'Ouchte et l'Irtich, Aug. 20, 1895, Chaff."–Danguy, l.c.: Qinhe district, Nun'tai village, 2600 m, No. 979, Aug. 6, 1957–Ching), Tien Shan (Sairam lake, 2450 m, July 12; Sairam, Talkibash mountain range, July 14 and 19–1877, A. Reg.; nor trail of Talki mountain range, on south-east. bank of Sairam-Nor, 2150 m, Aug. 18, 1953–Mois.; nor. bank of Sairam lake, 2800 m, No. 2124, Aug. 28, 1957–Shen), *Jung. Gobi* ("Baitag-Bogdo mountain range, No. 2709, Aug. 1, 1988"–Gub., Kamelin, Ganbold et al).

? IIIB. Tibet: *Chang Tang* ("on Sanju Pass in Jarkand, at 15,000 ft"–Henders. et Hume, l.c.; "33°39' lat., 82°45'57" long., 16,000 ft, Aug. 20, 1896"–Deasy, l.c.; "in 91°40', 35°21', 16,812 ft, Aug. 12, 1896"–Wellby et Malcolm; "Getszi"–Fl. Xizang, l.c.), *South.* ("Nimu, Linzhou"–Fl. Xizang, l.c.).

IIIC. Pamir (head of Chon-arek river, mossy tundra near glacier, 5000–6000 m, July 23, 1942–Serp.).

General distribution: Jung.-Tarb., Nor. and Cent. Tien Shan, East. Pam.; Europe, Mid. Asia (Talassk. Ala Tau), West. Sib. (Altay), East. Sib., Nor. Mong. (Fore Hubs., Hent., Hang.).

Note. 1. This species has been reckoned in "Flora Tibeta" [Flora of Tibet] (Fl. Xizang, l.c. 1986) but cited only for Sinkiang in "Flora Kitaya" [Flora of China] (Fl. Sin. l.c. 1988!). I did not study this species from Tibet and am not confident about the veracity of the data presented.

2. Occasionally, plants with 4-merous flowers treated even by N. Turczaninow as *G. tenella* Rottb. f. *tetramera* Turcz. (Fl. baic.-dahur. 2 (1856) 251) along with type f. *pentamera* Turcz. l.c., are found all over the distribution range of this species. Such forms have been pointed out for other species of this section as well and there is no justification to treat them as independent species.

## Section 9. Endotricha Froel.

73. G. acuta Michx. Fl. Bor. Amer. 1 (1803) 177; Grossh. in Fl. SSSR, 18 (1952) 607; Hilbig et Schamsran in Feddes Repert. 91 (1980) 39; Grub. Opred. rast. Mong. [Key to Plants of Mongolia] (1982) 201.–*G. amarella* L.s. l.: Kryl. Fl. Zap. Sib. 9 (1937) 2172, p.p.; Kitag. Lin. Fl. Mansh. (1939)

358; Grub. Konsp. fl. MNR [Conspectus of Flora of Mongolian People's Republic] (1955) 223; Hilbig et Schamsran, l.c. — *Gentianella acuta* (Michx.) Hulten in Mem. Soc. Fauna et Fl. Fenn. 25 (1950) 76; Fl. Intramong. 5 (1980) 81; Pl. vasc. Helanshan (1986) 203; Fl. Sin. 62 (1988) 318.

—Ic.: Fl. Intramong. 5, tab. 34, fig. 5–10.

Described from Nor. America. Type in Paris (P).

In wet and coastal meadows, along banks of brooks and around springs, forest borders and in larch forests, under shade of rocks and in scrubs.

IA. Mongolia: *Khobd.* (Umne-Otor-Ula [Kharkhira], 30 km nor.-east of Khobdo somon centre, nor. trail, July 31, 1977—Karam., Sanczir et al; "Tal d. K l. Charchiraa oberh. Erholungsheim, Waldlichtung; Echen-Chargant, Suslan Chamar, Waldlichtung" – Hilbig et Schamsran, l.c.), *Mong. Alt.* (on one of Taishir-ola mountains, July 16, 1894 — Klem.), *East. Mong.* (Mongolia chinensis in reditu e China, 1842 — Kirilow; "Datsinshan', Mankhanshan', Ulashan'"—Fl. Intramong. l.c.), *Gobi-Alt.* (Ikhe-Bogdo-Ula, nor. slope, upper belt, Aug. 23, 1926 — Tug.; Dundu-Saikhan, Aug. 17; Dzun-Saikhan, Aug. 22; Dzun-Saikhan, Yolo creek valley, July 23–1931, Ik.-Gal.; Dundu and Dzun-Saikhan mountain ranges, July–Aug. 1933 — M. Simukova), *Alash. Gobi* (Alashan mountain range: "Helanshan"—Pl. vasc. Helanshan, l.c., Fl. Intramong. l.c.).

General distribution: East. Sib., Far East (Okhot.), Nor. Mong., China (Dunbei, Nor., Nor.-West.), Nor. America.

## Section 10. Arctophila Griseb.

84

74. G. arenaria Maxim. Diagn. pl. nov. asiat. 8 (1893) 30.—*Gentianella arenaria* (Maxim.) T.N. Ho in Acta Biol. Plateau Sin. 1 (1982) 39; Fl. Xizang, 3 (1986) 973; Fl. Sin. 62 (1988) 315.—*Gentiana thomsonii* auct. non Clarke: Hemsl. Fl. Tibet (1902) 191.—*G. saposhnikovii* Pachom. in Consp. fl. Asiae Mediae, 8 (1986) 170 and 49, syn. nov.

—Ic.: Fl. Xizang, 3, tab. 369, fig. 5–6; Fl. Sin. 62, tab. 51, fig. 7–9.

Described from Tibet (Weitzan). Type in St.-Petersburg (LE). Plate III, fig. 3; map 5.

In wet and marshy meadows, along banks of lakes and brooks, on shoals high in mountains, 3400–5400 m.

IIIA. Qinghai: *Nanshan* (Sulei-khe river, 3650–3950 m, July 24; Yamatyn-umru mountains, 3950–4250 m, July 27–1894, Rob.; South. Kukunor mountain range, Tszagasyten-ketel' pass, under pass, 3550 m, Aug. 17, 1901 — Lad.).

IIIB. Tibet: *Chang Tang* (Przewalsky mountain range, nor. slope, 4250 m, Aug. 1890 — Rob.), *Weitzan* (Ripa S. lacus Expeditionis, 13,500 ft. in arena frequens, July 25– Aug. 5, 1884, No. 356—Przew., typus !; Yantszy-tszyan river basin, head of I-chyu river: Rkhombomtso lake, 4000 m, Aug. 1, 1900 — Lad.; Yantszy-tszyan river basin, slopes of Puchekh-la pass, 3950 m, under pass, July 29, 1900 — Lad.).

General distribution: Cent. Tien Shan (Sarydzhas river); China (Nor.-West.: Gansu).

75. G. azurea Bge. in Nouv. Mem. Soc. natur. Moscou, 7 (1829) 230; Griseb. in DC. Prodr. 9 (1845) 98; Ldb. Fl. Ross. 3 (1847) 57; Kryl. Fl. Zap. Sib. 9 (1937) 2176; Grossh. in Fl. SSSR, 18 (1952) 613; Grub; Konsp. fl. MNR [Conspectus of Flora of Mongolian People's Republic] (1955) 223; Fl. Kirgiz. 8 (1959) 195; Ikonnik. Opred. rast. Pamira [Key to Plants of Pamir] (1963) 198; Fl. Kazakhst. 7 (1964) 112; Grub. Opred. rast. Mong. [Key to Plants of Mongolia] (1982) 202; Pakhom. in Opred. rast. Sr. Azii [Key to Plants of Mid. Asia] 8 (1986) 48. — *Gentianella azurea* (Bge.) Holub in Folia Geobot. et Phytotax. (Praha) 2 (1957) 116, in adnot.; Fl. Xizang, 3 (1986) 978; Fl. Sin. 62 (1988) 317. — *Gentiana aurea* auct. non Bge.: Clarke in Hook. f. Fl. Brit. India, 4 (1883) 108.

—Ic.: Fl. Sin. 62, tab. 51, fig. 10–12; Bge. l.c. tab. 10, fig. 3.

Described from Siberia. Type in St.-Petersburg (LE).

In wet meadows, syrts (elevated watersheds), rocky mountain tundra, on talus, along banks of lakes and brooks, in *Cobresia*, larch and spruce groves, 2300–5000 m.

IA. Mongolia: *Mong. Alt.* (Dain-gol lake, bank of south-east. bay, July 12, 1908 — Sap.; nor. slope of Khara-Adzarga mountain range, vicinity of Khairkhan-Duru, Aug. 25, 1930 — Pob.; Khasagtu-Khairkhan mountain range, nor. slope of Tsagan-Irmyk-ul, in upper course of Khunkerin-ama, about 2500 m, July 23, 1972 — Grub., Dariima et al.; Gichgeniin-nuru 40 km east-nor.-east of Tsogt somon centre, 3100 m, Aug. 12, 1973 — Isach. et Rachk.), *Gobi-Alt.* (Ikhe-Bogdo mountains, nor. slope, upper belt, Aug. 23, 1926 — Tug.; Dzun-Saikhan mountains, top of Ëlo creek valley, Aug. 23, 1931 — Ik.-Gal.; Dundu- and Dzun-Saikhan mountain ranges, slopes and gorges, up to upper belt, July-Aug. 1933 — M. Simukova; Ikhe-Bogdo mountain range, south. slope, Sept. 6, 1943 — Yun.).

IIA. Junggar: *Tien Shan* (head of Algoi, 2750–3050 m, Sept. 11, 1880 — A. Reg.; near Bedel' pass, 1886 — Krasnov; Barchat, north of Yakou, No. 1746, Aug. 31, 1957 — Shen; from Nilki to Dzin-kho, through Yakou, No. 4039, 4076, Sept. 1, 1957 — Kuan; Mal. Luitsichen in Khomote, 3020 m, No. 7154, Aug. 9, 1958 — Lee et Chu).

IIIA. Qinghai: *Nanshan* (Tetung river above Yunan'-chen' settlement, Aug. 8, 1890 — Gr.-Grzh.; Ulan-bulak area, June 17, 1894; nor. slope of Humboldt mountain range, Chon-sai gorge, near Tangyl-khutul' pass, about 3650 m, July 23, 1895 — Rob.; South. Kukunor mountain range, Tszagasyten-ketel' pass, 3400–3550 m, Aug. 17, 1901 — Lad.; Mon'yuan', at head of Ganshig river, tributary of Peishikhe river, 3900–4300 m, Aug. 18, 1958 — Dolgushin).

IIIB. Tibet: *Weitzan* (Yantszy-tszyan river basin, along valley of Nko-gun brook entering Rkhombo-mtso lake, 4050 m, Aug. 6; same site, Darindo area near Chzherku monastery, 3400 m, Aug. 8; same site, vicinity of Chzherku monastery, 3400 m, Aug. 16 — 1900, Lad.; "Bizhu" — Fl. Xizang, l.c.).

IIIC. Pamir (Goo-dzhiro river, 4500–5500 m, July 27, 1942 — Serp.).

General distribution: Jung. Alat., Nor. and Cent. Tien Shan, East. Pam.; West. Sib. (Altay), East. Sib. (Sayans, Daur.), Nor. Mong. (Hent., Hang., Fore Hubs. ?), China (Nor.-West.; Gansu, South-West.: Sichuan, Yunnan).

76. G. moorcroftiana Wall. ex Griseb. in DC. Prodr. 9 (1845) 96, p.p.; Clarke in Hook. f. Fl. Brit. India, 4 (1893) 108. — *G. moorcroftiana* var. *maddenii* Clarke, l.c. 108; Strachey, Catal. (1906) 114. — *Gentianella*

*moorcroftiana* (Wall. ex Griseb.) Airy-Shaw in Hook. Ic. 35 (1943) tab. 3431, in adnot.; Hara, Chater et Williams, Enum. flow. pl. Nepal, 3 (1982) 94; Fl. Xizang, 3 (1986) 974; Fl. Sin. 62 (1988) 317.— *G. maddenii* (Clarke) Airy-Shaw, l.c.; Hara, Chater et Williams, l.c. 94.

—Ic.: Fl. Xizang, 3, tab. 368, fig. 7–12.

Described from Himalayas (Kashmir). Type in London (K).

Along meadowy slopes in alpine belt, up to 4500 m.

IIIB. Tibet: *South.* (Ali: "Pulan'"—Fl. Xizang, l.c.).

General distribution: Himalayas (west. Kashmir).

77. G. pygmaea Rgl. et Schmalh. in Izv. obshch. lyubit. estestvozn., antrop., etnogr. 34, 2 (1882) 54; Grossh. in Fl. SSSR, 18 (1952) 619; Fl. Kirgiz. 8 (1959) 196; Ikonnik. Opred. rast. Pamira [Key to Plants of Pamir] (1963) 199; Pakhom. in Opred. rast. Sr. Azii [Key to Plants of Mid. Asia] 8 (1986) 48.— *G. thomsonii* Clarke in Hook. f. Fl. Brit. India, 4 (1883) 109.— *Gentianella pygmaea* (Rgl. et Schmalh.) H. Smith apud Nilsson in Grana palynol. 7 (1967) 106 and 144; Ikonnik. in Novit. syst. pl. vasc. 6 (1969) 270; Fl. Xizang, 3 (1986) 973; Fl. Sin. 62 (1988) 316.

—Ic.: Fl. SSSR, 18, Plate 32, fig. 1.

Described from East. Pamir. Type in St.-Petersburg (LE).

In alpine meadows, moraines, rocky slopes, along banks of brooks, on pebble beds, 1700–5300 m.

IIIA. Qinghai: *Nanshan* (Kansu, Aug. 3, 1890—Martin; Yamatyn-umru mountains, 3950–4250 m, July 27, 1894, No. 364—Rob.).

IIIB. Tibet: *Chang Tang* (Keri mountain range, Chivei area, alpine belt, 3650 m, on brook, July 15, 1885—Przew.; "Getszi"—Fl. Xizang, l.c.), *Weitzan* (Yantszy-tszyan river basin, vicinity of Rkhombo-mtso lake, on marsh, Aug. 1, 1900—Lad.), *South.* ("Lhasa, Chzhada"—Fl. Xizang, l.c.).

IIIC. Pamir (Ulug-tuz gorge in Charlym river basin, at upper juniper boundary, June 22; Chicheklik river valley, along descents, July 28–1909, Divn.).

General distribution: Cent. Tien Shan, East. Pam.; China (South-West.: nor. Sichuan).

78. G. turkestanorum Gand. in Bull. Soc. Bot. France, 65 (1918) 60; Grossh. in Fl. SSSR, 18 (1952) 610; Grub. Konsp. fl. MNR [Conspectus of Flora of Mongolian People's Republic] (1955) 225; Fl. Kirgiz. 8 (1959) 194; Ikonnik. Opred. rast. Pamira [Key to Plants of Pamir] (1963) 198; Fl. Kazakhst. 7 (1964) 111; Grub. Opred. rast. Mong. [Key to Plants of Mongolia] (1982) 202; Pakhom. in Opred. rast. Sr. Azii [Key to Plants of Mid. Asia] 8 (1986) 47.— *G. umbellata* var. *glomerata* Kryl. Fl. sib. occid. 9 (1937) 2176.— *G. aurea* auct. non L.: Persson in Bot. notiser (1938) 297. — *Gentianella turkestanorum* (Gand.) Holub in Folia geobot. et phytotax. (Praha) 2 (1967) 118; Fl. Sin. 62 (1988) 313.

—Ic.: Fl. Kazakhst. 7, Plate 13, fig. 3; Fl. Sin. 62, tab. 51, fig. 1–3.

Described from Mid. Asia. Type in Paris (P).

86     In wet meadows, along banks of rivers and brooks, scrubs, spruce and larch forests and their borders, under shade of rocks, 1200–3100 m.

IA. Mongolia: *Mong. Alt.* (bank of Dzhirgalante river flowing between Narin and Shadzagai rivers, July 20, 1898 — Klem.; Dain-gol lake, slopes, July 5, 1906 — Sap.; valley of Urtu-gol river, nor. slope, Aug. 17, Khara-Adzarga mountain range, valley of Sakhir-sala river, along east. slope of Imertsik mountains, Aug. 22; same site, mountain slopes, Aug. 25; same site, around Khairkhan-Duru river, Aug. 25; valley of Khoit-Ulyasutai river, along bank of dry river bed, Aug. 30; Khara-Adzarga mountain range, rocky bank of Naituren-gol river, Sept. 2–1930, Pob.; midcourse of Bidzhiin-gol river, 25–30 km south of Tamchi-daba pass, left bank of valley slope, birch grove near spring, Aug. 10, 1947 — Yun.).

IB. Kashgar: *West.* (Tokhta-khon mountains, nor. slope, 3050–3650 m, in alpine meadows, July 21 and 25, 1889 — Rob.; valley of Sulu-Sokal 25 km from Irkeshtam, 2800–2900 m, July 26, 1935 — N. Olsuf'ev).

IIA. Junggar: *Tarb.* (north of Dachen town, 2250 m, No. 1577, Aug. 13–1957, Shen; same site, No. 2926, Aug. 13, 1957 — Kuan), *Jung. Alat.* (mountains in Toli district, in Albakzin region, No. 2584, Aug. 7; in Ala Tau mountains, in forest, No. 4686, Aug. 7–1957, Kuan; mountains in Toli town region [Barktok-Arba-kezen'], No. 1321, Aug. 7; same site, 1780 m, No. 1364, No. 1373, Aug. 8–1957, Shen; 20 km south of Ven'tsyuan' town, 2810 m, No. 1488, Aug. 14; vicinity of Ven'tsyuan' town, along river bank. No. 1813, Aug. 22–1957, Shen), *Tien Shan* (many reports throughout the region), *Jung. Gobi* ("Baitag-Bogdo mountain range" — Grub. 1982, l.c.).

IIIC. Pamir (Chaili-bulun area on Pas-rabat river, on brook, Aug. 2, 1909 — Divn.; Tashkurgan, July 25, 1913 — O. Knorring; "Jerzil, 3200 m, July 16, 1930; Bostan-terek, about 3000 m, Aug. 7 and about 2400 m, Aug. 8, 1934" — Persson; 0.5 km below Keinz-Agzy-Chad village along right tributary of Mia river, 3500 m, July 20; Mia river basin, 2300 m, July 22–1941; Pakhpu river gorge, 2700 m, Aug. 2, 1942 — Serp.).

General distribution: Jung.-Tarb., Nor. and Cent. Tien Shan, East. Pam.; West. Sib. (Altay), Mid. Asia (West. Tien Shan, Pam.-Alay).

Note. Colour of corolla is highly variable from dark blue through light blue to lilac, pink and light golden yellow. Length of corolla varies greatly from 8 to 19 mm.

## Section 11. Cyclostigma Griseb.

79. G. uniflora Georgi, Bemerk. Reise, 1 (1775) 204, non Willd. 1797; Bobr. in Not. syst. (Leningrad) 20 (1960) 12; Grub. Opred. rast. Mong. [Key to Plants of Mongolia] (1982) 201; Pakhom. in Opred. rast. Sr. Azii [Key to Plants of Mid. Asia] 8 (1986) 45. — *G. verna* β *alata* Ldb. Fl. Ross. 3 (1847) 61. — *G. verna* α *angulosa* Kryl. Fl. Alt. 3 (1904) 855; Kusn. in Acta Horti Petrop. 15 (1904) 451. — *G. angulosa* auct. non MB. 1808; Turcz. Fl. baic.-dahur. 2, 1 (1854) 248; Kryl. Fl. Zap. Sib. 9 (2193). — *G. krylovii* Grossh. in Dokl. AN Azerb SSR, 3 (1947) 32; Grossh. in Fl. SSSR, 18 (1952) 585; Grub. Konsp. fl. MNR [Conspectus of Flora of Mongolian People's Republic] (1955) 224; Fl. Kirgiz. 8 (1959) 193; Fl. Kazakhst. 7 (1964) 108.

—Ic.: Georgi, l.c. tab. 11, fig. 1; Fl. Kazakhst. 7, Plate 12, fig. 10; Grub. Opred. rast. Mong. [Key to Plants of Mongolia] Plate 108, fig. 485.

99

Described from East. Siberia (Trans-Baikal region). Type–cited sketch of Georgi. Map 2.

In alpine grasslands, sasas (solonchaks on lateral seepages), turf-covered rock screes and talus, moraines, 2100-3500 m.

IA. Mongolia: *Khobd.* (upper south. head of Kharkhira river along nor. slope of valley, July 23, 1879—Pot.; valley of Turgen' river, July 6, 1973—Banzragch, Karam et al), *Mong. Alt.* (Dain-gol lake, June 28, 1903—Gr.-Grzh.; Tsagan-gol river, Prokhodnaya river gorge, June 30, 1905—Sap.).

IIA. Junggar: *Tarb.* (Saur mountain range, south. slope, valley of Karagaitu river, creek valley on right bank of Bayan-Tsagan, June 23, 1957—Yun., Lee, Yuan'), *Jung. Alat.* (along Kazan river in upper Khorgos, 2100 m, June 22, 1878—A. Reg.), *Tien Shan* (Talki-bash mountain range, July 20; Sairam lake, 2450 m, July 20, 1877; Sumbe pass, 2100-2450 m, June 21, 1878—A. Reg.).

General distribution: Jung.-Tarb., Nor. and Cent. Tien Shan; West. Sib. (Altay, East. Sib. (Sayans, Baikal region, Daur.), Nor. Mong. (Fore Hubs., Hang.).

## 4. Lomatogonium A. Br.

in Flora (Regensb.) 13 (1830) 221; *Pleurogyne* Eschsch. in Linnaea, 1 (1826) 187, pro syn. sect. *Gentiana*; Benth. et Hook. f. Gen. pl. 2 (1876) 816, pro gen.

1. Flowers wide open, corolla stellate; during anthesis, lobes of calyx horizontally procumbent (section 1. Lomatogonium) .... 2.
+ Flowers cup-shaped; during anthesis, lobes of calyx erect (section 2. Pleurogynella (Ikonnik.) T.N. Ho) .......................... 7.
2. Leaves ovoid, obovoid, elliptical to oblong-ovoid and broad-lanceolate; lobes of calyx 1/3-1/2 of corolla .......................... 3.
+ Leaves linear or linear-lanceolate; lobes of calyx linear and nearly as long as corolla .......................... 6. L. rotatum (L.) Fries.
3. Lobes of calyx spatulate, obtuse or ovoid, cuspidate, with orbicular notches between them. Leaves ovoid or obovoid ..... 4.
+ Lobes of calyx lanceolate, cuspidate and notches between them acute. Leaves lanceolate or oblong-ovoid .......................... 5.
4. Lobes of calyx spatulate with orbicular tip. Lobes of corolla 4-8 mm long. Leaves elliptical, cuspidate ..........................
.......................... 2. L. chumbicum (Burk.) H. Smith.
+ Lobes of calyx ovoid, cuspidate, shortly tapered at base. Lobes of corolla 6-12 mm long. Leaves ovoid-spatulate with orbicular tip .......................... 3. L. gamosepalum (Burk.) H. Smith.
5. Flowers large; lobes of corolla 13-20 mm long; lobes of calyx narrow-lanceolate. Stem few-branched, with rather few flowers at tip and in axils of ovoid leaves ..........................
.......................... 5. L. macranthum (Diels et Gilg) Fern.

+ Flowers small; lobes of corolla 8–13 mm long. Stem branched from base ................................................................. 6.

6. Leaves and lobes of calyx oblong-ovoid; lobes of corolla 7–8 mm long (up to 10–11 mm long during fruiting). Plant 5–10 (15) cm tall ....................................... 1. L. carinthiacum (Wulf.) Reichb.

+ Leaves and lobes of calyx lanceolate; lobes of corolla 10–13 mm long. Plant 14–28 cm tall ..... 4. L. lloydioides (Burk.) H. Smith.

7. Lobes of corolla white, with longitudinal, nearly black band outside; lobes of calyx acute, oblong-lanceolate, without sacciform intumescence at base, rarely with faintly manifest intumescence ...................... 7. L. brachyantherum (Clarke) Fern.

+ Lobes of corolla bright or dark blue, without black band on back; lobes of calyx obtuse, oblong-ovoid, with sacciform intumescence at base ...................... 8. L. thomsonii (Clarke) Fern.

88

# Section 1. Lomatogonium

1. L. carinthiacum (Wulf.) Reichb. Fl. Germ. Excurs. (1831) 421; Kryl. Fl. Zap. Sib. 9 (1937) 2195; Persson in Bot. notiser (1938) 297; Bobr. in Fl. SSSR, 18 (1952) 620; Grub. Konsp. fl. MNR [Conspectus of Flora of Mongolian People's Republic] (1955) 225; Fl. Kirgiz. 8 (1959) 197; Ikonnik. Opred. rast. Pamira [Key to Plants of Pamir] (1963) 199; Fl. Kazakhst. 7 (1964) 114; Grub. Opred. rast. Mong. [Key to Plants of Mongolia] (1982) 203; Pakhom. in Opred. rast. Sr. Azii [Key to Plants of Mid. Asia] 8 (1986) 50; Fl. Xizang, 3 (1986) 980; Fl. Sin 62 (1988) 333. — *Swertia carinthiaca* Wulf. in Jacq. Misc. bot. 2 (1781) 53. — *Pleurogyne carinthiaca* Griseb. Gen. et sp. Gentian. (1838) 310; Clarke in Hook. f. Fl. Brit. India, 4 (1883) 120; Diels in Futterer, Durch Asien, 3 (1903) 18; Gr.-Grzh. Zap. Kitai, 3 (1907) 494; Diels in Filchner, Wissensch. Ergebn. 10, 2 (1908) 262; Danguy in Bull. Mus. nat. hist. natur. 20 (1914) 76; Hao in Bot. Jahrb. 68 (1938) 630.

— Ic.: Fl. SSSR, 18, Plate 33, fig. 2; Grub. Opred. rast. Mong. [Key to Plants of Mongolia] Plate 108, fig. 488; Fl. Sin. 62, tab. 54, fig. 5–7.

Described from Europe. Type in Munich (M).

In alpine and coastal meadows, *Cobresia* thickets, mountain spruce and larch forests, under rocks, in moraine, 800–5400 m.

IA. Mongolia: *Khobd.* (south. slope of Gurban-Khara-Ula 7 km west of Urkhai-Suren mine, Aug. 2, 1977—Karam., Sanczir et al), *Mong. Alt.* (along Sair-Byuira river, affluent of Dain-Gol lake, Sept. 25, 1876—Pot.*; Khashatu river, Aug. 8–9, 1899—Lad.; valley of Sagsai river near Nikiforovka trading station. Aug. 1, 1909—Sap.; Khara-Adzarga mountain range around Khairkhan-Duru river, Aug. 25, 1930—Pob.; Khasagtu-Khairkhan mountain range, nor. slope of Tsagan-Irmyk-Ula in upper Khunkerin-Ama, 2450–2500 m, Aug. 23, 1972—Grub., Ulzij. et al; Adzhi-Bogdo mountain range, 3300 m,

Aug. 23, 1973 – Rachk.), *Cent. Khalkha* (Ubur-Dzhargalante river between its head and Agit mountain, Aug. 31 and Sept. 2; left bank of Ubur-Dzhargalante river between Botaga and Agit mountains, subalpine zone, Sept. 11–1925, Krasch.; vicinity of Ikhe-Tukhum-Nur lake, June 1926 – Polynov).

IIA. Junggar: *Cis-Alt.* ("entre l'Ouchte et l'Irtich, 900 m, Aug. 20–1895, Chaff." – Danguy, l.c.), *Jung. Alat.* (south. slope below Koketau pass, July 21, 1909 – Lipsky), *Tien Shan* (south. bank of Sairam lake, 1877; Muzart valley, 2150–2450 m, Aug. 17; same site, below pass, 2750–3200 m, Aug. 19, 1877; Urtaksary river, 1850 m, Aug. 4; Dzhagastai south of Kul'dzha, 2150 m, Aug. 10; nor.-east of Sairam lake, 2150–2450 m, Aug. 18; upper Borotala, 1850 m and 2600 m, Aug.; Chubaty pass, 2750 m, July–1878; Mengete, 2750 m, Aug. 2; same site, 3050–3350 m, July 4; Zagastai-Gol, 2750 m, Sept. 5–1879, A. Reg.; Urten-Muzart gorge, Aug. 3, 1877 – Fet.; east of Sarytyur pass, Kotsai area, 2450 m, July 22, 1893 – Rob.; oberstes Agias Tal, Aug. 11–20; beim Hauptlager im Tal Khaptu-su, Sept. 1–5, 1907; am Wege von Fucan zur Bogdo-Ola, Aug. 2–3, 1908 – Merzb.; "Julduz, about 2240 m, Aug. 11, 1932" – Persson, l.c.; on south-east. bank of Sairam lake, about 2150 m, Aug. 18, 1953 – Mois.; nor. bank of Sairam lake, 2800 m, No. 2122, Aug. 27; Barchat, nor. Yakou, No. 1945, Aug. 31–1957, Shen; from Nilki to Dzinkho, No. 4067, Sept. 1, 1957 – Kuan; 15 km south of Tyan'chi lake, No. 1948, Sept. 19, 1957 – Shen; from Bortu to timber plant in Khomote, 2460 m, No. 7025, Aug. 4; in Khomote, 3020 m, No. 7157, Aug. 9; Aksu district, west of Oi-Terek, 2750 m, in spruce grove, No. 8293, Sept. 10–1958, Lee et Chu).

IIIA. Qinghai: *Nanshan* (North. Tetung mountain range, Cherik pass, 4000 m, Aug. 8, 1890 – Gr.-Grzh.; Yamatyn-Umru mountains, Yamatyn-Umru area, 3650 m, July 21; same site, 3650–3950 m, July 22; South. Kukunor mountain range, Kuku-Boguchi pass, 4250 m, July 24–1894; nor. slope of Humboldt mountain range, Chan-sai gorge near pass, about 3650 m, July 23, 1895 – Rob.; South. Kukunor mountain range, Tszagasyten-kotel pass, 3600–3650 m, Aug. 17, 1901 – Lad.; "am südlichen Ufer des Kuke-nur, Futterer" – Diels, l.c.), *Amdo* ("Kokonor, in der Nähe von Tsigigan-ba, um 3340 m, 1930" – Hao, l.c.).

IIIB. Tibet: *Weitzan* (nor. slope of Burkhan-Budda mountain range, 3950–4725 m, Aug. 13; same site, Khatu-Gol pass, 3100 m, Aug. 27–1884, Przew.; same site, Khatu gorge, 3200 m, July 25 and 29, 1901 – Lad.; Golubaya river basin, head of I-chyu river, vicinity of Rkhombo-mtso lake, 3980 m, Aug. 6; Darindo area near Chzherku monastery, 3560 m, Aug. 8; waterdivide of Mekong and Golubaya river basins, Go-chyu brook, 3950–4000 m, nor. slope of pass, Aug. 23; nor. descents of Gur-la pass, 4250 m, Aug. 24–1900, Lad.), *Chang Tang* (Keri mountain range, Syren-Bulun river, 3350 m, Aug. 3, 1885 – Przew.; Tokhta-Khon mountains, nor. slope, 3050–3650 m, July 21, 1889 – Rob.; "Bizhu" – Fl. Xizang, l.c.), *South.* ("Mochzhungunka" – Fl. Xizang, l.c.).

IIIC. Pamir (around Muzaling pass, July 30, 1909 – Divn.; Karatash area, 3500–4000 m, Sept. 8, 1941 – Serp.).

General distribution: Jung.-Tarb., Nor. and Cent. Tien Shan, East. Pam.; Europe (cent.), Caucasus, Mid. Asia (West. Tien Shan, Pam.-Alay), West. Sib. (Altay), East. Sib. (south), Far East (Kamchatka), Nor. Mong., China (Nor., Nor.-West., South-West.).

Note. K.I. Maximowicz identified the variety with 4-merous flowers – var. *tetramera* Maxim. (asterisked).

2. L. chumbicum (Burk.) H. Smith apud Nilsson in Grana palyn. 7, 1 (1967) 109, 145; Fl. Xizang, 3 (1986) 979; Fl. Sin. 62 (1988) 331. – *Swertia chumbica* Burk. in J. Asiat. Soc. Bengal, nov. ser. 2 (1906) 323.

– Ic.: Fl. Xizang, 3, tab. 369 fig. 1–2; Fl. Sin. 62, tab. 54, fig. 1–4.

Described from East. Himalayas (Yatung). Type in London (K).

On meadowy slopes in high mountains, 3500–4700 m.

IIIB. Tibet: *South.* ("Lankatsza" — Fl. Xizang, l.c.).

General distribution: Himalayas (east.).

3. L. gamosepalum (Burk.) H. Smith apud Nilsson in Grana palyn. 7, 1 (1967) 109, 145; Fl. Xizang, 3 (1986) 978; Fl. Sin. 62 (1988) 329. — *Swertia gamosepala* Burk. in J. Asiat. Soc. Bengal, nov. ser. 2 (1906) 324. — *Pleurogyne gamosepala* Hao in Bot. Jahrb. 68 (1938) 630.

— Ic.: Fl. Sin. 62, tab. 55, fig. 1–4.

Described from South-West. China (Sichuan). Type in Paris (P).

In alpine meadows, standing river shoals, among shrubs, in undergrowth and forest borders, 2800–4500 m.

IIIA. Qinghai: *Nanshan* ("Kokonor, 3000 m, zwischen steppen-pflanzen, 1930" — Hao, l.c.).

IIIB. Tibet: *Weitzan* ("Batsin" — Fl. Xizang, l.c.).

General distribution: China (South-West.: Kam, Sichuan, south. Gansu), Himalayas (east.).

4. L. lloydioides (Burk) H. Smith ex Chater in Hara, Chater et Williams, Enum. flow. pl. Nepal, 3 (1982) 95; Fl. Xizang, 3 (1936) 979; Fl. Sin. 62 (1988) 333. — *Swertia lloydioides* Burk. in J. Asiat. Soc. Bengal, nov. ser. 2 (1906) 323.

Described from South. Tibet. Type in London (K).

On rubbly and rocky mountain slopes, 4100–4500 m.

IIIB. Tibet: *South.* ("ad castrum Khamba-jong, Prain, 1637" typus ! — Burk. l.c.; "Mochzhungunka" — Fl. Xizang, l.c.).

General distribution: China (South-West.: Kam).

5. L. macranthum (Diels et Gilg) Fern. in Rhodora, 21 (1919) 197; Fl. Xizang, 3 (1986) 980; Fl. Sin. 62 (1988) 334. — *Pleurogyne macrantha* Diels et Gilg in Futterer, Durch Asien, 3 (1903) 17. — *Swertia deltoidea* Burk. in J. Asiat. Soc. Bengal, nov. ser. 2 (1906) 324. — *Pleurogyne carinthiaca* var. *bella* (Hemsl.) Diels: Hao in Bot. Jahrb. 68 (1938) 630.

— Ic.: Diels et Gilg, l.c. tab. 2; Fl. Sin. 62, tab. 55, fig. 5–8.

90    Described from Qinghai (Gunkhe). Type in Berlin (B).

On meadowy slopes of mountains, coastal meadows and shoals, among undergrowth in thin forests, alpine meadows, 2500–4800 m.

IIIA. Qinghai: *Nanshan* ("Schalakuto über der Wasserscheide zum Küke-nur", "bei Lager XXI am Nordfuss der Semenowberge" — Diels et Gilg, l.c.; "Kokonor auf dem Selgen, 4800 m, 1930" — Hao, l.c.), *Amdo* ("zwischen Lager XXV und XXVI, östlich vom Hoang-ho, am Nordfuss des Dschupargebirges" — Diels et Gilg, l.c.).

IIIB. Tibet: South. ("Lankatsza" — Fl. Xizang, l.c.).

General distribution: Himalayas (east.).

6. L. rotatum (L.) Fries ex Nym. Consp. fl. Europ. (1881) 500; Kryl. Fl. Zap. Sib. 9 (1937) 2195; Bobr. in Fl. SSSR, 18 (1952) 621; Grub. Konsp. fl. MNR [Conspectus of Flora of Mongolian People's Republic] (1955) 226; Fl. Kazakhst. 7 (1964) 114; Fl. Intramong. 5 (1980) 85; Grub. Opred. rast.

Mong. [Key to Plants of Mongolia] (1982) 204; Pakhom. in Opred rast. Sr. Azii [Key to Plants of Mid. Asia] 8 (1986) 50; Fl. Sin. 62 (1988) 336.— *Swertia rotata* L. Sp. pl. (1753) 226.— *Pleurogyne rotata* Griseb. Gen. et sp. Gentian. (1838) 309; Forbes et Hemsl. Index Fl. Sin. 2 (1890) 138; Danguy in Bull. Mus. nat. hist. natur. 20 (1914) 76.

—Ic.: Fl. SSSR, 18, Plate 33, fig. 3; Fl. Intramong. 5, tab. 36, fig. 6–10; Fl. Sin. 62, tab. 55, fig. 9–12.

Described from East. Siberia. Type in London (Linn.).

In moist, marshy, sometimes solonetzic meadows and grassy marshes, along banks of rivers and lakes, on pebble beds, 1000–4200 m.

IA. Mongolia: *Khobd.* (vicinity of Achit-Nur), Mong. Alt. (on Ugumyn river, Sept. 22, 1899—Lad.; Bombotu-Khairkhan mountains, on bank of Khapchik river, Oct. 10, 1930— Pob.), *Cent. Khalkha* (central Kerulen near Tsagan-Obo mountain; same site, near Dalai-Beise duke's site—1890, Pal.; Ubur-Dzhargalante river between its head and Agit mountain, Aug. 30, 1925—Krasch. et Zam.; vicinity of Ikhe-Tukhum-Nur, June 1926; bank of Khaldangin-Nor lake along road from Mishik-Gun to Dol'che-Gegen, June 1926 —Zam.), *East. Mong.* ("south. Ulantsab ajmaq—administrative territorial unit in Mongolia—Fl. Intramong. l.c.), *Depr. Lakes* (along bank of Ol'ge-Nur lake, Aug. 29; Ulangom, Sept. 6–1879, Pot.; Shargain-gobi around Gol-Ikhe, Sept. 4 1930—Pob.; nor. bank of Khirgiz-Nur lake near Chizyrgany-bulak spring, Aug. 21, 1944—Yun.), *Ordos* (on Bayan-Nor lake, Sept. 2; Gundzhagatai area, on lake, Sept. 7–1884, Pot.).

IIA. Junggar: Tien Shan (B. Yuldus, 2450 m, Aug. 5, 1893; Tor-kul' lake [Turkul'], Aug. 25, 1895—Rob.; Barkul' lake, No. 4500, Sept. 25, 1957—Kuan; same site ? No. 2151, Sept. 25, 1957—Shen; B. Yuldus basin, 30–35 km west—south-west of Bain-Bulak settlement, Aug. 10, 1958—Yun., I.-f. Yuan').

IIIA. Qinghai: *Nanshan* (South. Tetung mountain range, Sept. 28, 1872; Kuku-Usu river, 3050 m, Aug. 6, 1879—Przew.; North. Tetung mountain range, Cherik pass, 4000 m, Aug. 8, 1890—Gr.-Grzh.; Kuku-Nor lake, Tsungu-Dzhara area, Sept. 11, 1908—Czet.; South. Kukunor mountain range, nor. slope, Tsaiza-Gol, 3650 m, Sept. 9, 1894—Rob.; same site, around Tszagasyten-Këtel pass, 3600–3650 m, Aug. 17, 1901—Lad.).

IIIB. Tibet: *Weitzan* (waterdivide of Golubaya river and Mekong river basin, nor. slope of Gurla pass, 4250 m, Aug. 24, 1900—Lad.).

General distribution: Jung-Tarb., Nor. Tien Shan; Arct. (Europ.), West. Sib. (Altay), East. Sib. (south.), Far East, Nor. Mong., China (Dunbei, Nor., Nor.-West., South-West.), Japan, Nor. Amer. (nor.).

## Section 2. Pleurogynella (Ikonnik.) T.N. Ho

7. L. brachyantherum (Clarke) Fern. in Rhodora, 21 (1919) 197; Fl. Xizang, 3 (1986) 979; Fl. Sin. 62 (1988) 339.— *Pleurogyne brachyanthera* Clarke in Hook. f. Fl. Brit. India, 4 (1883) 120; Deasy, In Tibet and Chin. Turk. (1901) 398; Hemsl. Fl. Tibet (1902) 191; Hedin, S. Tibet, June 3 (1922) 48; Ikonnik. Opred. rast. Pamira [Key to Plants of Pamir] (1963) 200.— *P. diffusa* Maxim. in Bull. Ac. Sci. St.-Petersb. 32 (1888) 510; Hemsl. in J. Linn. Soc. London (Bot.) 30 (1894) 117; Deasy, In Tibet and Chin. Turk. (1903) 403.— *P. carinthiaca* auct. non Griseb.: Hemsl. et Pears. in Peterm. Mitt. 28 (1900) 374.— *Pleurogynella brachyanthera* (Clarke) Ikonnik.

in Novit. syst. pl. vasc. (1969) 271; Pakhom. in Opred. rast. Sr. Azii [Key to Plants of Mid. Asia] 8 (1986) 50.

Described from Kashmir. Type in London (K). Plate II, fig. 2.

In wet alpine meadows, along banks of brooks and rivers, among shrubs, 3000–5200 m.

IIIA. Qinghai: *Nanshan* (along Kuku-Usu river, 3050 m, Aug. 6, 1879 — Przew.; Yamatyn-Umru mountain range, alpine meadow, 3650–4250 m, July 21; same site, 3650–3950 m, July 22–1894, Rob.; nor. slope of Humboldt mountain range, Chon-sai gorge near pass, about 3650 m, Aug. 23, 1895 — Rob.).

IIIB. Tibet: *Weitzan* (nor. slope of Burkhan-Budda mountain range, 3100 m, among shrubs, Aug. 15 [27], 1884 — Przew., typus *Pleurogyne diffusa* Maxim. !), *Chang Tang* ("Tsaidam, hill-side close to water at 5150 m — Thoreld, 1891; "on the south-east. shore of Aru-tso, 5200 m, Aug. 5, 1896 — Deasy et Pike" — Hemsl. l.c.; "Camp 30: lat. 33°48′10″, long. 82°29′60″, 5200 m, Aug. 5, 1896" — Deasy, l.c.; "Kwenlun, Sarik-kol, 3469 m, Aug. 5, 1896" — Hedin, l.c.; "Aksu, 4790 m, 1898" — Deasy, l.c.).

IIIC. Pamir (Kok-Muinak pass, on sand along bank of brook, July 27; Chicheklik valley, in wet meadow, July 28 — 1909, Divn.).

General distribution: Cent. Tien Shan (Sarydzhaz river basin, Aksay valley), East. Pam.; Mid. Asia (Badakhshan), China (Kam), Himalayas.

8. L. thomsonii (Clarke) Fern. in Rhodora, 21 (1919) 197; Fl. Xizang, 3 (1986) 979; Fl. Sin. 62 (1988) 340. — *Pleurogyne thomsonii* Clarke, in Hook. f. Fl. Brit. India, 4 (1883) 120; Alcock, Rep. natur. hist. results Pamir bound. commiss. (1898) 24; Ikonnik. Opred. rast. Pamira [Key to Plants of Pamir] (1963) 200. — *Pleurogynella thomsonii* (Clarke) Ikonnik. in Novit. syst. pl. vasc. (1969) 271; Pakhom. in Opred. rast. Sr. Azii [Key to Plants of Mid. Asia] 8 (1986) 50.

— Ic.: Fl. Sin. 62, tab. 53, fig. 9–12.

Described from Kashmir (Karakorum). Type in London (K).

In alpine meadows, shoals, coastal and marshy meadows, 2200-5200 m.

IIIA. Qinghai: *Nanshan* (road from Alashan to Kuku-Nor lake, Sangyn area, up to upper belt, on all slopes, Aug. 12, 1908 — Czet.).

IIIB. Tibet: *Weitzan* (basin of Golubaya river, head of I-chyu, vicinity of Rkhombo-mtso lake, 4000 m, Aug. 6, 1900 — Lad.), *Chang Tang* ("Gaitsze, Getszy, Dan-syun, Zhitu"), *South.* ('Pulan', Shigatsze" — Fl. Xizang, l.c.).

IIIC. Pamir (?).

General distribution: East. Pam.; China (Nor.-West.: Gansu), Himalayas.

## 5. Lomatogoniopsis T.N. Ho et S.W. Liu.
### in Acta Phytotax. sin. 18, 4 (1980) 466

1. L. alpina T.N. Ho et S.W. Liu, l.c. 467; Fl. Xizang, 3 (1986) 981; Fl. Sin. 62 (1988) 341.

—Ic.: T.N. Ho et S.W. Liu, l.c. 468, tab. 1, fig. 1–5; Fl. Sin. 62, tab. 56, fig. 1–4.

Described from Tibet (Weitzan). Type in Xining (HNWP). Plate IV, fig. 3.

92    In alpine meadows and among shrubs, 4000–4300 m.

IIIB. Tibet: *Weitzan* ("Zadoi, on meadow, alt. 4200 m, Aug. 23, 1965, No. 511—S.W. Liu, typus !: Jigzhi, under bushes, alt. 4300 m, Aug. 19, 1971, No. 588; same loc., on meadow, alt. 3950 m, Sept. 1, 1971, No. 633—Qinghai Inst. Biol. Geolog. Exped." —T.N. Ho et S.W. Liu, l.c.).

General distribution: China (Sikan).

## 6. Anagallidium Griseb.
### Gen. et Sp. Gentian. (1838) 311

1. A. dichotomum (L.) Griseb. Gen. et Sp. Gentian. (1838) 312; Franch. Pl. David. 1 (1884) 212; Kryl. Fl. Zap. Sib. 9 (1937) 2196; Kitag. Lin. Fl. Mansh. (1939) 357; Grossh. in Fl. SSSR, 18 (1952) 622; Grub. Konsp. fl. MNR [Conspectus of Flora of Mongolian People's Republic] (1955) 226; Fl. Kirgiz. 8 (1959) 197; Fl. Kazakhst. 7 (1964) 116; Grub. Opred. rast. Mong. [Key to Plants of Mongolia] (1982) 203; Pakhom. in Opred. rast. Sr. Azii [Key to Plants of Mid. Asia] 8 (1986) 51.—*Swertia dichotoma* L. Sp. pl. (1753) 227; Danguy in Bull. Mus. nat. hist. natur. 20 (1914) 76; Fl. Intramong. 5 (1980) 87; Pl. vasc. Helanshan (1986) 203; Fl. Sin. 62 (1988) 404.

—Ic.: Fl. Kazakhst. 7, Plate 14, fig. 5; Fl. Intramong. 5, tab. 36, fig. 1–5.

Described from Siberia. Type in London (Linn.).

In wet coastal and forest meadows, river shoals and coastal pebble beds, thickets of coastal shrubs.

IA. Mongolia: *Cent. Khalkha* (basin of Dzhargalante river, 47° N. lat., 104–105° E. long. between head of Ubur-Dzhargalante river and Agit mountain, Aug. 10, Aug. 31, Sept. 2 and Sept. 4, 1925—Krasch. et Zam.), *East. Mong.* (Muni-Ula mountains, July 8, 1871—Przew.; "Vallee du Keroulen, 1896, Chaff."—Danguy, l.c.; vicinity of Khailar town, June 13, 1951—Sh.-s. Lee), *Alash. Gobi* (Alashan mountain range, Yamata gorge, May 5, 1908—Czet.).

IIA. Junggar: *Tien Shan* (Muzart, Aug. 1877—A. Reg.; Tekes, Muzart river, 3350 m, July 3; on Tekes river and on nor. slopes of mountains, 3350–3950 m, July 7–1893, Rob.; 8 km east of Chzhaosu, 1740 m, No. 785, Aug. 11; on Chzhaosu-Shaty road, No. 831, Aug. 12,—1957, Shen; in Savan region, 1100 m, No. 1790, June 23, 1957—Kuan).

IIIA. Qinghai: *Nanshan* (Mon'yuan', valley of Tetung-khe near stud farm, 2800 m, nor.-east. slopes, Aug. 20, 1958—L. Dolgushin), *Amdo* ("Mudzhik mountains, 3050 m, along river under coastal scarps, June 17, 1880—Przew.).

General distribution: Fore Balkh. (Zaisan basin), Jung.-Tarb., Nor. Tien Shan; West. Sib. (south-east Altay), East. Sib. (south.), Nor. Mong., China (Dunbei, Nor., Nor.-West., Cent.).

## 7. Swertia L.
### Sp. pl. (1753) 226

1. Perennials with simple stem branched only in inflorescence and well-distinguished radical rosette of large leaves on more or less long petioles (section 1. Swertia) ........................................ 2.

+ Annuals with stem often branched from base and poorly developed radical rosette of short-petiolate leaves perishing by anthesis (section 2. Ophelia (D. Don) Benth. et Hook. f.) ....... 10.

2. Flowers nutant; corolla yellow-green with red-brown speckles, its lobes with a single large orbicular brown nectary at base. Cauline leaves developed, ovoid, opposite ..................................... ................................................................ 4. S. erythrosticta Maxim.

+ Flowers erect; corolla of a different colour, its lobes with 2 colourless nectaries at base. Cauline leaves poorly developed, opposite or alternate, lanceolate or oblong, rather few or absent but for a pair under inflorescence .............................................. 3.

3. Cauline leaves alternate, only uppermost 2 leaflets supporting inflorescence often opposite; radical and lower cauline leaves on long petiole, nearly equal to elliptical or ovoid blade ......... 4.

+ All cauline leaves opposite or absent but for a pair of small ones under inflorescence; stem without leaves .................................... 5.

4. Corolla grey-violet or dark blue, its lobes obtuse, entire ............ ................................................................................. 6. S. obtusa Ldb.

+ Corolla yellow-green with violet dots and streaks inside; its lobes obtuse, crenulate or emarginate at tip ..................................... ................................................................ 1. S. banzragczii Sancz.

5. Large plant with thick, up to 1 cm in diam., stem, 60–100 cm tall. Cauline leaves joined together into tall (1–3 cm) sheath; radical leaves elliptical or broad-elliptical, 15–35 cm long, with 5–7 nerves, quite sharply tapered into long petiole. Flowers small; corolla yellowish white, with violet dots and streaks inside, its lobes 7–10 mm long ................... 3. S. connata Schrenk.

+ Stem 10–45 cm tall. Cauline leaves free or absent; radical leaves small, from obovoid and oval to linear, gradually tapered into short petiole. Flowers much larger; lobes of corolla 10–30 mm long .......................................................................................... 6.

6. Stem without leaves; developed leaves only in radical rosette. Inflorescence few-flowered ........................................................ 7.

+ Apart from rosetted leaves, 3–4 pairs of cauline leaves developed. Flowers large (lobes of corolla 10–30 mm long), 1–3 each in axillary umbels in many-flowered inflorescence; corolla

yellow. Leaves narrow-lanceolate to linear, 3–8 mm broad and 2.5–6 cm long; radical leaves erect ...................................................
...................................................... 9. S. younghusbandii Burk.

7. Radical rosette consists of 1–2 pairs of obovoid or oval leaves. Lobes of corolla 15–22 mm long, 5–8 mm broad ...................... 8.

+ Radical rosette consists of 3–4 pairs of very narrow, oblong-oval or lanceolate leaves. Lobes of corolla 10–12 mm long, 2–3 mm broad, light or dark blue with broad white border .....................
.................................................................. 5. S. marginata Schrenk.

8. Corolla light blue. Radical leaves only one pair ...........................
.............................................................. 2. S. bifolia Batal.

+ Corolla yellow-green, light blue only outside on back of lobes. Radical leaves 1–2 pairs ................................................................. 9.

9. Lobes of corolla 9–15 mm long, lanceolate, acuminate. Ovary rugose ...................................... 7. S. przewalskii Pissjauk.

+ Lobes of corolla 15–20 mm long, oval, obtuse and dentate at tip. Ovary glabrous ............................ 8. S. wolfgangiana Grüning.

10. Flowers 4-merous, 2–3 times larger in terminal inflorescence (lobes of corolla 9–12 mm long) than on slender branches in lower part of stem; corolla yellow-green, sometimes with lilac band on back of lobes; nectary single in the form of obovoid cavity covered laterally by lobed spathes unevenly .....................
.................................................... 15. S. tetraptera Maxim.

+ Flowers 5-merous, rarely 4-merous; all of them identical; nectaries fimbriate or ciliate, paired, rarely single .................. 11.

11. Stamens fused at base into short tube; nectary single, crescent-shaped, with a pair of dark spots above upper rim. Flowers 5-merous; corolla lilac-coloured, its lobes declinate during anthesis ...................................... 10. S. ciliata (D. Don) B.L. Burtt.

+ Stamens free. Nectaries paired,without spots ........................ 12.

12. Leaves and lobes of calyx with short and stiff cilia along margins. Nectaries pocket-shaped, open downward toward base of lobes of corolla and short-fimbriate along margin. Corolla light lilac or white. Leaves linear-oblong, obtuse ..........
.............................................................. 13. S. hispidicalyx Burk.

+ Leaves and lobes of calyx glabrous. Nectaries open upward toward tip of lobes of corolla. Leaves acute ............................ 13.

13. Leaves linear or linear-lanceolate. Lobes of calyx equal to or slightly longer than lobes of corolla, subulate. Flowers 5-merous, 10–15 mm in diam.; corolla light blue; nectaries pocket-shaped, long-fimbriate ..........................................................
.......................................... 11. S. diluta (Turcz.) Benth. et Hook. f.

+ Leaves ovoid-lanceolate or narrow-deltoid-cordate, semiamplexicaul at base. Calyx 1/2–2/3 of corolla, its lobes lanceolate. Nectaries orbicular, ovoid or oblong, open upward ................ 14.

14. Flowers 4-merous; corolla red-lilac, sometimes yellow, 8–13 mm in diam.; nectaries sulcate, oblong, short-ciliate along margins ........................................................................ 14. S. mussotii Franch.

+ Flowers 5-merous, occasionally 4-merous; corolla light blue or white; nectaries depressed, orbicular or ovoid, fimbriate ....... 15.

15. Flowers 15–25 mm in diam., 5-merous; corolla light blue; nectaries large, ovoid, long- and thick-fimbriate. Leaves narrow-deltoid-cordate ................ 12. S. franchetiana H. Smith.

+ Flowers 4–8 mm in diam., 5-merous, occasionally 4-merous; corolla light blue or white; nectaries small, orbicular, short- and sparse-fimbriate. Leaves ovoid–lanceolate.....................................
...................................................S. macrosperma (Clarke) Clarke.

## Section 1. Swertia

1. S. banzragczii Sanczir in Novit. syst. pl. vasc. 21 (1984) 136—*S. obtusa* auct. non Ldb.: Grub. Konsp. fl. MNR [Conspectus of Flora of Mongolian People's Republic] (1955) 226, p.p.; id. Opred. rast. Mong. [Key to Plants of Mongolia] (1982) 223, p.p.

Described from Mongolia (Mong. Altay). Type in St.-Petersburg (LE). Map 4.

In alpine meadows, dwarf birch groves, sparse larch groves and their borders in alpine belt, 1800–3000 m.

IA. Mongolia: *Mong. Alt.* (mountain slopes on upper Ulyasty river, July 22; mountain slopes on Naryn river upper courses, July 23–1898, Klem.; Kutologoi eastward of Ak-Korum pass, alpine tundra, July 13, 1908—Sap.; basin of Bulgan river, upper Indertiin-gol, Kharagaitin-daba pass, July 24; upper Ketsu-Sairin-Gol, along moraine and slopes, July 26–1947, Yun.; basin of Buyantu river, Chigirtei-gol 12 km above same-named lake, nor. slope of Chigirtei-ul, about 2800 m, young larch grove, July 4, 1971—Grub., Dariima et al; Khara-Us-nur lake, Zhold-Ama area, Aug. 6, 1976—Beket, paratypus !; in dwarf birch groves along southern slope of Dayan-Nur lake, July 26, 1977—Sanczir, Karam. et al, typus !; basin of Sanginin-gol 25 km south-east of Dayan-nur camp, 1800–2000 m, July 16–17; 10–15 km from Dayan-nur camp near Khara-nur lake, July 20; basin of Ikhe-Dzhargalant-gol [right tributary of Bulgan river] 10–15 km north-west of Bulgan settlement, July 27–1988, Kam., Ganbold et al), Khobd. ?

IIA. Junggar: Cis-Alt. (Qinhe [Chingil'], 1800 m, No. 953, Aug. 4; between Nun'tai and Chzhunkhaitsza, 2500 m, No. 1155, Aug. 5–1956, Ching).

General distribution: East. Sib. (West. Sayan: reported by author on Shapshal mountain range in Tuva).

2. S. bifolia Batal. in Acta Horti Petrop. 13 (1894) 378; Diels in Futterer, Durch Asien (1903) 18; Hao in Bot. Jahrb. 68 (1938) 630; Fl.

Xizang, 3 (1986) 985; Fl. Sin. 62 (1988) 362. — *S. wardii* Marq. in J. Linn. Soc. London (Bot.) 48 (1929) 208; ? S. sp. Walker in Contribs U.S. Nat. Herb. 28 (1941) 653.

—Ic.: Fl. Xizang, 3, tab. 370, Fl. Sin. 62, tab. 58, fig. 1–4.

Described from South-West. China (Sichuan). Type in St.-Petersburg (LE).

On alpine meadowy slopes, wet, marshy and shrub-covered meadows, on borders and under canopy of forest, 2850–4500 m.

IIIA. Qinghai: *Nanshan* ("am südlichen Ufer des Küke-nur"–Diels, l.c.; "an den südabhängen der süd-kokonorischen Gebirge Selgen und Hoto-Selgen, 3800 m, 1930" — Hao, l.c.; "La Chi Tzu Shan, on partially shaded, very moist steppes, No. 706, 1923, Ching. Flowers turquoise blue" — Walker, l.c. ?).

General distribution: China (Nor.-West.: south. Gansu, mountainous Shenxi; South-West.).

3. S. connata Schrenk in Fisch. et Mey. Enum. pl. nov. 1 (1841) 37, 2 (1842) 31; Ldb. Fl. Ross. 3 (1847) 75; Grossh. in Fl. SSSR, 18 (1952) 632; Fl. Kirgiz. 8 (1959) 199; Fl. Kazakhst. 7 (1964) 117; Pakhom. in Opred. rast. Sr. Azii [Key to Plants of Mid. Asia] 8 (1986) 52; Fl. Sin. 62 (1988) 358.

Described from East. Kazakhstan (Jung. Ala Tau). Type in St.-Petersburg (LE).

In wet and marshy meadows, along banks of brooks, in willow thickets, borders and clearances in spruce forests, 1600–2650 m.

IIA. Junggar: *Jung. Alat.* {[around Kazan pass ?] 2100–2450 m, ? Aug. 1878 — A. Reg.), *Tien Shan* (Talki brook, July 20, 1877 — A. Reg.; Sairam lake, July 10; 33 versts [1 verst = 1.067 km] — from Sairam lake, July 18–1878, Fet.; Chubaty pass in Sairam region, 2450–2750 m, Aug. 2, 1878; Borgaty, 1500–1850 m, July 4; Turgun-tsagan, July; Aryslyn, 2450 m, July 10; same site, July 18; same site, 2450–2750 m, July 19; lower Aryslyn, 1680–1850 m, July 20; Dzhirgalan nor.-east of Kul'dzha, July 26; Mengute, 2750 m, Aug. 2; Mengute pass, 3050–3350 m, Aug. — 1879, A. Reg.; 20 km nor. of Ulastai, No. 3871, Aug. 28, 1957 — Kuan; Narat mountains in Kunges, 2300 m, No. 6570, Aug. 7, 1958 — Lee et Chu; nor. foothills of Narat mountain range in Tsanma valley, 2250 m, Aug. 7, 1958 — Yun., Yuan').

General distribution: Jung.-Alat., Nor. and Cent. Tien Shan.

4. S. erythrosticta Maxim. in Bull. Ac. Sci. St.-Petersb. 27 (1881) 503; Forbes et Hemsl. Index Fl. Sin. 2 (1890) 140; Fl. Sin. 62 (1988) 355.

—Ic.: Fl. Sin. 62, tab. 57, fig. 1–5.

Described from Qinghai. Type in St.-Petersburg (LE).

Along meadowy and meadowy-steppe slopes, river shoals and coastal meadows, sparse forests, 1500–4300 m.

IIIA. Qinghai: *Nanshan* (Jugum S. a fl. Tetung, 7500 ft, in sylvis frondosis parce, July 26 [Aug. 7] 1880, No. 719 — Przew., typus).

General distribution: China (Dunbei nor., Nor.: Hebei, Nor.-West.: Gansu, Shanxi, South-West.: Sichuan; Cent. Hubei), Korean peninsula (nor.).

110

Note. This species cited in Fl. Sin. 62, l.c., for Inner Mongolia but not listed in Fl. Intramongolica.

5. S. marginata Schrenk in Bull. Ac. Sci. St.-Petersb. 10 (1842) 353; Ldb. Fl. Ross. 3 (1847) 75; Danguy in Bull. Mus. nat. hist. natur. 14 (1908) 131 and 20 (1914) 76 cum auct. Fisch. et Mey.; Persson in Bot. notiser (1938) 298; Grossh. in Fl. SSSR, 18 (1952) 635; Grub. Konsp. fl. MNR [Conspectus of Flora of Mongolian People's Republic] (1955) 226; Ikonnik. in Dokl. AN TadzhSSR, 20 (1957) 56; Fl. Kirgiz. 8 (1959) 199; Fl. Kazakhst. 7 (1964) 118; Grub. Opred. rast. Mong. [Key to Plants of Mongolia] (1982) 203; Pakhom. in Opred. rast. Sr. Azii [Key to Plants of Mid. Asia] 8 (1986) 52; Fl. Sin. 62 (1988) 364.—S. graciliflora Gontsch. in Acta Inst. bot. Ac. Sci. URSS, ser. 1, 1 (1933) 161; Grossh. in Fl. SSSR, 18 (1952) 636; Fl. Kirgiz. 8 (1959) 200; Fl. Tadzh. 7 (1984) 306; Pakhom. in Opred. rast. Sr. Azii [Key to Plants of Mid. Asia] 8 (1986) 53; Fl. Sin. 62 (1988) 364.—S. pseudopetiolata Pissjauk. in Not. syst. (Leningrad) 21 (1961) 295.—S. sp. Alcock, Rep. natur. hist. results Pamir bound. commiss. (1895) 25 ?

—Ic.: Fl. Kazakhst. 7, Plate 14, fig. 1; Grub. Opred. rast. Mong. [Key to plants of Mongolia] Plate 108, fig. 484; Fl. Sin. 62, tab. 58, fig. 11–14.

Described from East. Kazakhstan (Jung. Ala Tau). Type in St.-Petersburg (LE).

In wet meadows, syrts (elevated watersheds), along banks of brooks, in moraines, among shrubs, rubbly sites and slopes, talus and rocks, in spruce forests, 2500–4800 m.

IA. Mongolia: *Mong. Alt.* ("Altai, Mongolie, Sept. 1895, Chaff." —Danguy, l.c. Grub. l.c.) ?

IIA. Junggar: *Jung. Alat.* (Kazan pass, 2750–3350 m, Aug. 10, 1878—A. Reg.; 20 km south of Ven'tsyuan', 2810 m, No. 1475, Aug. 14, 1957—Shen), *Tien Shan* (Talkibash mountain range, July 20; Sairam lake, July 20–24; in Dzhagastai mountains, Aug. 2; Urten-Muzart gorge, Aug. 4; Muzart pass, 3500 m, Aug. 18; Muzart valley below pass, 2750–3200 m, Aug. 19; Muzart [foothills, 1500–1850 m], Aug. 20–1877; in Kok-kamyr mountains, 3050 m, July 25; Kyzemchek, 3050 m, July 3; Chubaty pass [on Sairam lake], 2450–2750 m, Aug. 2–1878; opposite Aryslan estuary, 2450 m, July 8; Aryslan on nor. side of Kash river, 2750–3050 m, July 12; Mengute, 2750 m, Aug. 2—typus *S. pseudopetiolata* Pissjauk. !; head of Kash river, 3050–3350 m, Aug. 12; Kash river, 2750 m, Aug. 17–18–1879, A. Reg.; east of Sary-tyur pass, 2900 m, Kok-sai river, July 22; Khaidyk-gol river, Seileng-tokhoi area, 2150 m, Aug. 6–1893, Rob.; Saryds-chass, nahe au Gletscherende, Aug. 2–8, 1903; Passe zwischen Kinsu und Kurdai, July 3, 1907—Merzb.; "Atchelek pass, about 2900 m, Aug. 9, 1932"—Persson, l.c.; Manas river basin, valley of Ulan-usu river 1–2 km above Dzhartas river estuary, July 18; basin of Manas river, upper Danu-gol at ascent to Se-daban pass, July 21–1957, Yun., Lee, Yuan'; Daban, 2300 m, No. 1973, July 18; between Danu and Daban, 3000 m, No. 2039, July 19; along Danu river, 2600 m, No. 2073, No. 2144, July 21–1957, Kuan; 36 km south of Nyutsyuan'tsza, No. 142, July 19; south of Danu, 3150 m, No. 400, July 21; 6–7 km south of Danu, No. 469, July 21; same site, No. 499, July 22–1957, Shen; Chzhaosu in Aksu region, 3000 m, No. 3518, Aug. 14; 10 km nor. of Chzhaosu, No. 3302, Aug. 15–1957, Kuan; ? No. 906, Aug. 15; Sarbushin, No. 2102, Aug. 26; 3 km south of Yakou [Barchat],

2900 m, No. 1669, Aug. 30–1957, Shen; Nilki district, 60 km north of Ulyastai, 2850 m, No. 3998. Aug. 31, 1957 — Kuan; on Urumchi-Yan'tsi highway, 2300 m, No. 5992, July 21, east of Khomote, Taskhan cannal, 300m, No. 7138, Aug. 6; Khesho district, Nyuitsygen, 3100 m, No. 7184, Aug. 10; Tsagan-nur pass in Khotun-Sumbul, 3100 m, No. 6536, Aug. 15–1958, Lee et Chu; Narat mountain range on road to Bain-Bulak settlement from Dachit pass, July 8; M. Yuldus, basin 7 km from Kotyl' pass on road to Karashar from Yuldus, Aug. 15–1958, Yun., Yuan').

97

IIIC. Pamir ("Mustagh-Ata, bords de la rivier Beik, 4200 m, July 23, 1906, Lacoste" — Danguy, 1908, l.c.; Muzaling pass, July 31, 1909 — Divn.; Arpalyk river, Kyzyl-daban, 3200 m, July 9, 1941; upper Kanlyk river, 4500–5000 m, July 14, 1942 — Serp.; "Kungur mountain range, south-west. slope, moraine of Kok-Sel' glacier, Yaman-Yarsu river, 4450 m, Aug. 8, 1956, Pen Shu-li et al" — Ikonnik. l.c.).

General distribution: Jung.-Alat., Nor. and Cent. Tien Shan. East. Pam.; Mid. Asia (West. Tien Shan, Pam.-Alay), East. Sib. (east. Sayan), Nor. Mong. (Fore Hubs.).

Note. It is impossible to distinguish *S. graciliflora* Gontsch. from *S. marginata* either from structural details and extent of spread of nectaries or from relative length of lobes of calyx and breadth of calyx margins. The above-cited characteristics fall within the range of natural variation of latter species, are unstable and are reported throughout the entire distribution range.

The differentiation of *S. pseudopetiolata* Pissjauk. from this species based on the number and form of radical leaves is untenable as these characteristics are unstable and depend on habitat conditions of their growth (soil fertility and moisture, daylight factor etc.) and age of plant.

The colour of corolla in *S. marginata* varies intensely throughout the distribution range from light to dark bluish or greyish violet.

6. S. obtusa Ldb. in Mem. Ac. Sci. St.-Petersb. 5 (1812) 526; ej. Fl. Ross. 3 (1847) 75; Kryl. Fl. Zap. Sib. 9 (1937) 2199; Grossh. in Fl. SSSR, 18 (1952) 631; Grub. Konsp. fl. MNR [Conspectus of Flora of Mongolian People's Republic] (1955) 226, p.p.; Fl. Kazakhst. 7 (1964) 117; Grub. Opred. rast. Mong. [Key to Plants of Mongolia] (1982) 203; p.p., Pakhom. in Opred. rast. Sr. Azii [Key to Plants of Mid. Asia] 8 (1986) 52; Fl. Sin. 62 (1988) 359. — S. perennis L. var. obtusa Griseb. Gen. et sp. Gentian. (1838) 331; Danguy in Bull. Mus. nat. hist. natur. 20 (1914) 76.

— Ic.: Fl. SSSR, 18, Plate 34, fig. 1; Fl. Kazakhst. 7 Plate 14, fig. 2.

Described from East. Siberia (Trans-Baikal). Type in St.-Petersburg (LE).

In meadows and alpine grasslands, along banks of brooks and streams, on rock screes and rocky slopes, 2100–2500 m.

IIA. Junggar: *Cis-Alt.* (along Oi-Chilik river, tributary of Kran river, Sept. 20, 1876 — Pot.; "entre l'Ouchte et l'Irtich, Aug. 27, 1895, Chaff." — Danguy, l.c.).

General distribution: Jung.-Tarb. (Saur., Tarb.); West. Sib. (Urals, Altay), East. Sib., Far East, China (Altay).

7. S. przewalskii Pissjauk. in Not. syst. (Leningrad) 21 (1961) 300; Fl. Sin 62 (1988) 362. — S. marginata auct. non Schrenk: Franch. in Bull. Soc. Bot. France, 46 (1899) 312, p.p. quoad pl. tangut.

— Ic.: Pissjauk. l.c. fig. 1 a–b; Fl. Sin. 62, tab. 58, fig. 5–10.

Described from Qinghai. Type in St.-Petersburg (LE). Plate II, fig. 3.

In alpine meadows, coastal meadows and scrubs, 2500–4200 m.

IIIA. Qinghai: *Nanshan* (South. Tetung mountain range, Aug. 27, 1872; Sodi-Soruksum mountain, 3950 m, July 31–Aug. 3, 1872; North. Tetung mountain range, 2450–2750 m, alpine meadow in middle forest belt, No. 814, Aug. 2 [14], 1880 — Przew., typus !; Chorten-ton temple, 2300–2700 m, Sept. 1901 — Lad.).

IIIB. Tibet: *Weitzan* (Yangtze river basin: Chamudug-la pass, 4800 m, along meadowy descents, July 26; Darindo area near Chzherku, Bounchin rocks, 3600 m, Aug. 8–9–1900, Lad.).

General distribution: endemic.

98    8. S. wolfgangiana Grüning in Feddes Repert. 12 (1913) 309; Fl. Sin. 62 (1988) 363. — *S. marginata* auct. non Schrenk: Hao in Bot. Jahrb. 68 (1938) 630.

Described from Nor.-West. China (Shanxi). Type in Berlin (B) ?

In moist, marshy and alpine meadows, shrubby meadowy slopes, along banks of rivers, 1500–5000 m.

IIIA. Qinghai: *Nanshan* ("auf dem Selgen; 4800 m, 1930" — Hao, l.c.), *Amdo* ("Amne-Matchin, in den Talern zwischen 4500–5000 m; auf dem Ming-ge-Gebirge, 4500 m, 1930" — Hao, l.c.).

General distribution: China (Nor.-West., South-West., Cent.: Hubei).

9. S. younghusbandii Burk. in J. Asiat. Soc. Bengal, nov. ser. 2 (1906) 325; Fl. Xizang, 3 (1986) 989; Fl. Sin. 62 (1988) 370.

—Ic.: Fl. Xizang, 3, tab. 371.

Described from South. Tibet. Type in Calcutta (Kolkota) (Cal).

In alpine meadows, among coastal shrubs, meadowy shrubby slopes of mountains, 4300–5400 m.

IIIB. Tibet: *South.* ("ad castrum Khamba-jong. ad 15,000 ped. Prain 1622, Younghusband 297" — typus ! — Burk. l.c.; "Nan'mulin" — Fl. Xizang, l.c.).

General distribution: China (South-West.: Kam), Himalayas (east.).

## Section 2. Ophelia (D. Don) Benth. et Hook. f.

10. S. ciliata (D. Don) Burtt in Not. Bot. Gard. Edinb. 26 (1965) 272, 277; Hara, Chater et Williams, Enum. flow. pl. Nepal, 3 (1982) 96; Fl. Xizang, 3 (1986) 995; Fl. Sin. 62 (1988) 407. — *Ophelia ciliata* D. Don ex G. Don, Gen. Syst. Dichlam. Pl. 4 (1837) 178. — *O. purpurascens* D. Don in Trans. Linn. Soc. 17 (1837) 526; Griseb. Gen. et Sp. Gentian. (1838) 315; id. in DC. Prodr. 9 (1845) 124. — *Swertia purpurascens* Wall. ex Clarke in Hook. f. Fl. Brit. India, 4 (1883) 121.

—Ic.: Fl. Xizang, 3, tab. 373.

Described from West. Himalayas. Type in London (K).

Along meadowy slopes and banks of brooks, 3600–3900 m.

IIIB. Tibet: *South.* ("Pulan'" — Fl. Xizang, l.c.).

General distribution: Himalayas (West. Kashmir, east.).

11. S. diluta (Turcz.) Benth. et Hook. f. Gen. pl. 2 (1876) 817; Kitag. Lin. Fl. Mansh. (1939) 360; Fl. Intramong. 5 (1980) 87; Fl. Sin. 62 (1988) 394. — *Sczukinia diluta* Turcz. in Bull. Soc. Natur. Moscou, 13 (1840) 166. — *Ophelia diluta* (Turcz.) Ldb. Fl. Ross. 3 (1847) 73; Franch. Pl. David. 1 (1884) 212; Kryl. Fl. Zap. Sib. 9 (1937) 2199; Grossh. in Fl. SSSR, 18 (1952) 626. — *O. chinensis* Bge. ex Griseb. in DC. Prodr. 9 (1845) 126; Franch. Pl. David. 1 (1884) 212; Grossh. in Fl. SSSR, 18 (1952) 625. — *Swertia chinensis* (Bge.) Franch. ex Hemsl. in J. Linn. Soc. London (Bot.) 26 (1890) 139; Franch. in Bull. Soc. Bot. France, 46 (1899) 322; Kitag. Lin. Fl. Mansh. (1939) 360.

—Ic.: Fl. Intramong. 5, tab. 37, fig. 9–11; Fl. Sin. 62, tab. 64, fig. 7–8.

Described from East. Siberia (Dauria). Type in St.-Petersburg (LE).

In wet and marshy meadows, along banks of rivers, in scrubs, up to 2600 m.

IA. Mongolia: *East. Mong.* (Shilin-Khoto, 1959—Ivan.; "Kulun-Buirnor. region, Ulashan'"—Fl. Intramong. l.c.), *Ordos* (on Bain-nor lake, Sept. 2; Gun'-dzhagatai area, Sept. 7–1884, Pot.), *Khesi* (Sutcheou, Sept. 10, 1890—Martin).

IIIA. Qinghai: *Nanshan* (North. Tetung mountain range, forest belt, 2300 m, Aug. 11, 1880—Przew.; in Itel'-gol river valley, March, 1885—Pot.; Kuku-Nor lake, Ui-yu area, Sept. 12, 1908; near Donkyr town, Aug. 5, 1909—Czet.).

General distribution: West. Sib. (east. Altay), East. Sib. (Sayans, Daur.), Far East (south.), China (Dunbei, Nor., Nor.-West., Cent., South-West.: nor. Sichuan).

12. S. franchetiana H. Smith in Bull. Brit. Mus. (Natur. Hist.) Bot. 4 (1970) 251; Fl. Xizang, 3 (1986) 994; Fl. Sin. 62 (1988) 402. — *S. stricta* Franch. in Bull. Soc. Bot. France, 46 (1899) 322, non Collett et Hemsl. 1890.

—Ic.: Fl. Xizang, 3, tab. 372, fig. 1–2; Fl. Sin. 62, tab 65, fig. 8–11.

Described from South-West. China (Sichuan). Type in Paris (P).

On meadowy slopes, river banks, ravine floors, among shrubs, forest borders, 2200–3700 m.

IIIA. Qinghai ? (Fl. Sin. l.c.).

IIIB. Tibet: *South.* ("Lhasa, Gunbutszyanda, Linzha"—Fl. Xizang, l.c.).

General distribution: China (Nor.-West.: south. Gansu, South-West.: Sichuan).

13. S. hispidicalyx Burk. in J. Asiat. Soc. Bengal, nov. ser. 2 (1906) 321; Hara, Chater et Williams, Enum. flow. pl. Nepal, 3 (1982) 97; Fl. Xizang, 3 (1986) 992; Fl. Sin. 62 (1988) 392.

—Ic.: Fl. Xizang, 3, tab. 372, fig. 7–8; Fl. Sin. 62, tab. 64, fig. 1–3.

Described from South. Tibet. Type in London (K) or in Calcutta (Kolkota) (Cal).

On meadowy slopes and alpine meadows, banks of rivers and brooks and lake basins, 3400–5200 m.

IIIB. Tibet: *Chang Tang* ("Gaitsze" — Fl. Xizang, l.c.), *South.* ("urbis Lhasa boream versus in faucibus Phemby-la dictis, Walton 1608 [typus ?] et orientem versus in valle fluminis Kyichu, Walton 1159; ad castrum Gyang-tze, Walton 1609; prope fines sikkimensis ad castrum Khamba-jong dictum, Younghusband 293" — Burk. l.c.; "Lhasa, Nan'mulin, Kanma, Pulan', Shigatsze" — Fl. Xizang, l.c.).

General distribution: Himalayas (east.).

S. macrosperma (Clarke) Clarke in Hook. f. Fl. Brit. India, 4 (1883) 123; H. Smith in Hand.-Mazz. Symb. Sin. 7 (1936) 987; Hara, Chater et Williams, Enum. flow. pl. Nepal, 3 (1982) 97; Fl. Sin. 62 (1988) 397. — *Ophelia macrosperma* Clarke in J. Linn. Soc. Bot. (London) 14 (1875) 448.

— Ic.: Fl. Sin. 62, tab. 65, fig. 1–3.

Described from India. Type in London (K).

In mountain meadows, along banks of rivers and brooks, among shrubs and in sparse forests, 1400–4000 m.

IIIB. Tibet (Fl. Sin. l.c.).

General distribution: China (South-West., Cent.: Hubei, South.: Guangxi, Taiwan), Himalayas (east.), Indo-Mal. (India, Burma).

Note. Not enumerated in Flora of Tibet (Fl. Xizang, 3 (1986)).

14. S. mussotii Franch. in Bull. Soc. Bot. France, 46 (1899) 316; H. Smith in Hand.-Mazz. Symb. Sin. 7 (1936) 987; Fl. Xizang, 3 (1986) 994; Fl. Sin. 62 (1988) 400.

— Ic: Fl. Sin. 62, tab. 65, fig. 4–7.

Described from South-West. China (Sichuan). Type in Paris (P).

On meadowy slopes, coastal meadows and shoals, in gorges, among shrubs, 2000–3800 m.

IIIA. Qinghai ("south-west. Qinghai" — Fl. Sin. l.c.).

IIIB. Tibet: *Weitzan* (basin of Yantszy-tszyan river, vicinity of Chzherku monastery, 3470 m, in ploughed fields, Aug. 11, 1900 — Lad.).

100    General distribution: China (South-West.).

15. S. tetraptera Maxim. in Bull. Ac. Sci. St.-Petersb. 27 (1881) 503; Forbes et Hemsl. Index Fl. Sin. 2 (1890) 141; Fl. Sin. 62 (1988) 405. — *S. dimorpha* Batal. in Acta Horti Petrop. 13 (1894) 379; Hao in Bot. Jahrb. 68 (1938) 630. — *S. pusila* Diels in Notizbl. Bot. Gart. u. Mus. Berlin, 11 (1931) 215, non Pursh, 1814; Walker in Contribs U.S. Nat. Herb. 28 (1941) 653.

–Ic.: Fl. Sin. 62, tab. 66, fig. 1–7.

Described from Qinghai. Type in St.-Petersburg (LE). Plate II, fig. 4.

In wet meadows, along banks of rivers, in scrubs, on meadowy slopes, 2000–4000 m.

IA. Mongolia: *Alash. Gobi* (Alashan mountain range; "Halahu-Kou, 2100 m, on moist valley bottoms, No. 70, May 1923, R.C. Ching" — Walker, l.c., Diels, l.c., typus *S. pusilla* Diels).

IIIA. Qinghai: *Nanshan* (South. Tetung mountain range, on Rangkhta-gol river, in valleys on south. slope in alpine belt, July 13, 1872−Przew.; "Kokonor: an den südabhängen des Selgen und Hoto-Selgen, 3800 m, 1930"−Hao, l.c.), *Amdo* (ad Fl. Mudshik-che, 9000–9500 ft ad ripam in fruticetis frequens, No. 404, June 17 [29] 1880−Przew., typus !).

IIIB. Tibet: *Weitzan* (basin of Yantszy-tszyan river, Donra area on Khi-chyu river, 3950 m, in scrubs along mountain slopes, July 16, 1900−Lad.).

General distribution: China (Nor.-West.: Gansu; South-West.: Sichuan).

## 8. Halenia Borkh.
## in Roem. Arch. Bot. 1, 1 (1796) 25

1. Flowers greenish pale yellow, 8–11 mm long; lobes of calyx lanceolate. Cauline leaves with 3 nerves, tapered at base. Plant 15–40 cm tall .................................... 1. H. corniculata (L.) Cornaz.

+ Flowers light blue, 4–5 mm long; lobes of calyx ovoid. Cauline leaves with 5 nerves, less tapered at base and subamplexicaul. Plant 40–60 cm tall .................................... 2. H. elliptica D. Don.

1. H. corniculata (L.). Cornaz in Bull. Soc. Sci. Nat. Neuch. 25 (1897) 171; Kryl. Fl. Zap. Sib. 9 (1937) 2198; Kitag. Lin. Fl. Mansh. (1939) 359; Grossh. in Fl. SSSR, 18 (1952) 638; Grub. Konsp. fl. MNR [Conspectus of Flora of Mongolian People's Republic] (1955) 226; Fl. Kazakhst. 7 (1964) 119; Grub. Opred. rast. Mong. [Key to Plants of Mongolia] (1982) 203; Fl. Sin. 62 (1988) 291.−*Swertia corniculata* L. Sp. pl. (1753) 227.−*Halenia sibirica* Borkh. in Roem. Arch. Bot. 1, 1 (1796) 25; Griseb. in DC. Prodr. 9 (1845) 128; Forbes et Hemsl. Index Fl. Sin. 2 (1890) 141; Diels in Futterer, Durch Asien (1903) 18; Fl. Intramong. 5 (1980) 91.

−Ic.: Fl. Intramong. 5, tab. 39, fig. 1–5; Fl. Sin. 62, tab. 48, fig. 1–2.

Described from Siberia. Type in London (Linn.).

In wet and coastal meadows, around springs, among coastal shrubs, sandy shoals, sink holes in steppes, forest borders, 200–1700 m.

IA. Mongolia: *East. Mong.* (Mongolia chinensis in reditu e China 1841, Kirilow; Shilin-Khoto, true steppe, 1960−Ivan.), *Khesi* (Sutcheou, Sept. 15, 1890−Martin).

IIIA. Qinghai: *Nanshan* ("Beim kloster Kumbum südlich Sining-fu, Futterer"−Diels, l.c.).

General distribution: Europe (nor.-west. Urals), West. Sib. (Altay and south-east.), East. Sib. (south.), Far East, Nor. Mong., China (Dunbei, Nor., Nor.-West), Korean peninsula, Japan.

101    2. H. elliptica D. Don in Trans. Linn. Soc. London, 27 (1837) 529; Griseb. in DC. Prodr. 9 (1845) 129; Clarke in Hook. f. Fl. Brit. India, 4 (1883) 130; Hao in Bot. Jahrb. 68 (1938) 630; Walker in Contribs U.S. Nat. Herb. 28 (1941) 653; Grossh. in Fl. SSSR, 18 (1952) 639; Fl. Kazakhst. 7 (1964) 119; Fl. Intramong. 5 (1980) 91; Pl. vasc. Helanshan (1986) 204;

Pakhom. in Opred. rast. Sr. Azii [Key to Plants of Mid. Asia] 8 (1986) 55; Fl. Sin. 62 (1988) 291.

—Ic.: Fl. SSSR, 18, Plate 34, fig. 2; Fl. Intramong. 5, tab. 39, fig. 6–10; Fl. Sin. 62, tab. 48, fig. 3–6.

Described from Himalayas. Type in London (Linn.). Plate II, fig. 1.

In wet, floodplain and forest meadows, along banks and shoals of rivers, borders of swamps, along irrigation ditches, among coastal shrubs, 700–4100 m.

IA. Mongolia: *Alash. Gobi* (Alashan: "Hola-Shan, on steppes and in open woods, 1923, R.C. Ching"—Walker, l.c.; Pl. vasc. Helanshan, l.c.).

IIA. Junggar: *Tien Shan* (Tekes river, 3350 m, July 10, 1893; Edir-gol river, Aug. 22, 1895—Rob.; 3 km nor. of Chzhaosu, No. 878, Aug. 13, 1957—Shen; same site, No. 3281, Aug. 13, 1957—Kuan; valley of upper Tekes 5–6 km from Aksu settlement on road to Kalmak-Kure, Aug. 25; floodplain of upper Tekes at its intersection with road to Kalmak-Kure from Aksu, Aug. 25–1957, Yun., Lee, Yuan').

IIIA. Qinghai: *Nanshan* (South. Tetung mountain range, along river bed, July 18; same site, in forest meadows, Aug. 7–1872, Przew.; Yusun-Khatyma river, July 24, 1880; without date, No. 706; without date, No. 443, 1880—Przew.; on Itel'-gol river, April 1885 —Pot.; Dangertin town, 2150 m, Aug. 24, 1901—Lad.; along bed of Pinfan-khe river, June 21; Tanfanza village, July 23–1908; along bed of Pinfan-khe river, July 21, 1909— Czet.; "auf dem östlichen Nan-Schan, 2900–3200 m"—Hao, l.c.; 25 km south of Gulan town, gentle slopes of low mountains, 2450 m, Aug. 12, 1958; 24 km south of Xining, 2650 m, slopes of knolls, Aug. 4, 1959—Petr.).

IIIB. Tibet: *South.* (Gyantse, Sept. 7, 1904—Walton).

General distribution: Nor. Tien Shan (east.); China (Dunbei south., Nor.-West., South-West., Cent.), Himalayas.

Note. Plants with lilac and white flowers are sometimes found.

## Family 97. MENYANTHACEAE Dum.

1. Leaves aerial, radical, ternate, with oval leaflets. Flowers white or pinkish, campanulate-infundibular, in terminal raceme on almost leafless stem. Capsule dehiscent on 2 valves .................... .................................................... 1. Menyanthes L. (*M. trifoliata* L.).

+ Leaves floating, alternate on long funiform stem, entire, orbicular-cordate. Flowers yellow, wide open, almost dish-shaped, in axillary clusters. Capsule indehiscent ......................... ............... 2. Nymphoides Seguier (*N. peltatum* (Gmel.) O. Ktze.).

## 1. Menyanthes L.
### Sp. pl. (1753) 145

1. M. trifoliata L. Sp. pl. (1753) 145; Griseb. Gen. et Sp. Gentian. (1838) 340; id. in DC. Prodr. 9 (1845) 137; Clarke in Hook. f. Fl. Brit. India, 4 (1883) 130; Danguy in Bull. Mus. nat. hist. natur. 20 (1914) 76; H. Smith

102 in Hand.-Mazz. Symb. Sin. 7 (1936) 988; Kitag. Lin. Fl. Mansh. (1939) 360; Bobr. in Fl. SSSR, 18 (1952) 642; Grub. Konsp. fl. MNR [Conspectus of Flora of Mongolian People's Republic] (1955) 226; Fl. Kazakhst. 7 (1964) 120; Ohwi, Fl. Jap. (1978) 1102; Hara, Chater et Williams, Enum. flow. pl. Nepal, 3 (1982) 91; Grub. Opred. rast. Mong. [Key to Plants of Mongolia] (1982) 203; Pakhom. in Opred. rast. Sr. Azii [Key to Plants of Mid. Asia] 8 (1986) 56; Fl. Sin. 62 (1988) 412.

—Ic.: Fl. Sin. 62, tab. 67, fig. 1–3.

Described from Europe. Type in London (Linn).

On sandy and marshy banks of rivers, meanders and lakes, in and on water.

IA. Mongolia: *East. Mong.* (right bank of Huang He river above Hekou town in flooded meadow, Aug. 3, 1884—Pot.; "Vallee du Keroulen, May 1896, Chaff."—Danguy, l.c.).

IIA. Junggar: *Zaisan* ?, *Dzhark.* ?

General distribution: Jung. Ala Tau; Arct. (Asian), Europe, Mediterr., Balk.-Asia Minor, Caucasus, West. and East. Sib., Far East, Nor. Mong. (Fore Hubs., Hent.), China (Altay), Himalayas, Nor. Amer.

Note. The report cited (Danguy, l.c.) does not state exact location or date of find and possibly refers to Mong. Daur. or Hent. regions of Nor. Mongolia. This species is also found in Zaisan and Dzharkent regions of Kazakhstan and may be found in adjoining Junggar.

## 2. Nymphoides Seguier
Pl. Veron. 3 (1754) 121.—*Limnanthemum* S.G. Gmel. in Novi Comm. Ac. Sci. Petrop. 14, 1 (1770) 527

1. N. peltatum (Gmel.) O. Ktze. Revis. Gen. pl. 2 (1891) 429; H. Smith in Hand.-Mazz. Symb. Sin. 7 (1936) 989; Kitag. Lin. Fl. Mansh. (1939) 360; Bobr. in Fl. SSSR, 18 (1952) 648; Grub. Konsp. fl. MNR [Conspectus of Flora of Mongolian People's Republic] (1955) 226; Fl. Kazakhst. 7 (1964) 122; Ohwi, Fl. Jap. (1978) 1103; Grub. Opred. rast. Mong. [Key to Plants of Mongolia] (1982) 203; Jager, Hanelt, Davazamc in Flora, 177 (1985) 72; Pakhom. in Opred. rast. Sr. Azii [Key to Plants of Mid. Asia] 8 (1986) 56; Fl. Sin. 62 (1988) 415; Gubanov, Kamelin, Ganbold et al. in Bot. zhurn. 74, 2 (1989) 264. —*Menyanthes nymphoides* L. Sp. pl. (1753) 145.—*Limnanthemum peltatum* S.G. Gmel. in Novi Comm. Ac. Sci. Petrop. 14, 1 (1770) 527. —*L. nymphoides* Hoffmgg. et Link, Fl. Portug. 1 (1809) 344; Griseb. Gen. et Sp. Gentian. (1889) 341; id. in DC. Prodr. 9 (1845) 138; Clarke in Hook. f. Fl. Brit. India, 4 (1883) 131; Forbes et Hemsl. Index Fl. Sin. 2 (1890) 142; Simpson in J. Linn. Soc. Bot. (London) 41 (1913) 433; Danguy in Bull. Mus. nat. hist. natur. 20 (1914) 76; Kryl. Fl. Zap. Sib. 9 (1937) 2202.

—Ic.: Fl. Sin. 62, tab. 67, fig. 4–6.

Described from Europe. Type in London (Linn).

In gently flowing rivers, backwaters, shallow lakes, meanders, swamps.

IA. Mongolia: *Cent. Khalkha* (Kerulen midcourse, in river valley near Batur-Chzhonon tszasak area, 1899—Pal.; Kerulen valley around Sergelen-gun monastery, Aug. 19; meanders of Kerulen left bank near Bain-Khan, Aug. 20–1924, Lis.; valley of Toly river in Ulkhuin-bulun area, Aug. 9, 1926—Gus.), *East. Mong.* (right bank of Huang He river above Hekou town, Aug. 7, 1884—Pot.; Kerulen river, June 8 and 9, 1899—Pot. et Sold.; Kerulen midcourse, valley of river above Bars-Khoto, 1899—Pal.; Ul'gein-gol river, July 25, 1899—Pot. et Sold.; Kerulen river 80 km above San-Beis, Aug. 15, 1928—Tug.; Buir-nur lake at confluence of Khalkhin-gol, July 13, 1965—U. Tsogt; Khalkhin-gol 10–40 km above Sumber settlement, July 15 and 25, 1985—Gub., Kam. et al), *Depr. Lakes* (Chon-Kharaikh river at Khara-Nur lake, Aug. 12; south. bank of Khara-Usu lake near Kobdo town, Aug. 16–1879, Pot.; same site, Aug. 28, 1899—Lad.; Khara-Usu lake, Edel'kan-Namchi, Aug. 14, 1931—Bar. et Shukhardin; south-east. coast of Khara-Nur lake, Aug. 27, 1972—Metel'tseva), *Ordos* (Huang He valley, Aug. 6, 1871—Przew.; 75 km from Dzhasakachi town, Tautykhai lake, Aug. 17, 1957—Petr.).

IB. Kashgar: *Nor.* (Kuruk-tag, Kurli river gorge near Bash-Akin picket, in stream, Aug. 24, 1929—Pop.).

IIA. Junggar: *Cis-Alt.* (Chern. Irtysh river [near Dyurbel'dzhin crossing], Aug. 26, 1876—Pot.; "Bords de l'Irtich, Aug. 30, 1895, Chaff."—Danguy, l.c.; "Stagnant backwaters of the Upper Irtish River, Price"—Simpson, l.c.), *Zaisan* (Chern. Irtysh river, left bank opposite Cherektas mountain, June 11, 1914—Shishkin), *Jung. Gobi* (nor.: Burchum river, Barbagai, No. 2902, Sept. 11, 1956—Ching; "Bulgan-gol Aue zwischen Gum-gol und Jarantaj, Alt-wasser und Buchten, July 1982, Jager"—Jager, Hanelt, Dawazamc, l.c.).

General distribution: Aralo-Casp., Fore Balkh., Jung.-Tarb.; Europe, Mediterr., Balk.-Asia Minor, Fore Asia, Caucasus, Mid. Asia (Plains), West. Sib., East. Sib. (south.), Far East (south.), Nor. Mong. (Mong.-Daur.), China, Himalayas, Japan.

# Family 98. APOCYNACEAE Juss.

## 1. Apocynum L.

Sp. pl. (1753) 213 and Gen. pl. ed. 5 (1754) 101.— *Trachomitum* Woodson in Ann. Missouri Bot. Gard. 17 (1930) 157. — *Poacynum* Baill. in Bull. Soc. Linn. Paris, 1 (1888) 757

1. Leaves opposite, sometimes, upper leaves alone alternate. Flowers 5–15 mm in diam., erect; corolla campanulate; ovary semi-inferior ...................................................... 1. A. venetum L.

+ Leaves alternate. Flowers 15–25 mm in diam., nutant; corolla dish-shaped; ovary superior .................... 2. A. pictum Schrenk.

1. A. venetum L. Sp. pl. (1753) 213; DC. Prodr. 8 (1844) 440; Ledeb. Fl. Ross. 3 (1846) 43; Diels in Futterer Durch Asien (1903) 18; Danguy in Bull. Mus. nat. hist. natur. 17 (1911) 339 and 20 (1914) 75; Persson in Bot. notiser (1938) 298; Kitag. Lin. Fl. Mansh. (1939) 361; Fl. Sin. 63 (1977) 158;

[Drev. rast. Tsinkhaya-Woody Plants of Qinghai] (1987) 353; Fl. Intramong. 4 (1993) 112. — *A. lancifolium* Russan. in Acta Inst. bot. Ac. Sci. URSS, 1 (1933) 167; Kryl. Fl. Zap. Sib. 9 (1937) 2206. — *Trachomitum venetum* (L.) Woodson in Ann. Missouri Bot. Gard. 17 (1930) 158; Walker in Contribs U.S. Nat. Herb. 28 (1941) 653; Chen et Chou, Rast. pokrov dol. r. Sulekhe [Vegetational cover of Sulekhe River Valley] (1957) 89. — *T. lancifolium* (Russan.) Pobed. in Fl. URSS, 18 (1952) 658; Grub. in Bot. mat. (Leningrad) 19 (1959) 548; Fl. Kirgiz. 8 (1959) 204; Fl. Kazakhst. 7 (1964) 124; Butkov in Opred. rast. Sr. Azii [Key to Plants of Mid. Asia] 8 (1986) 57; Jager, Hanelt et Davazamc in Flora, 177 (1985) 32.

—Ic.: Fl. Sin. 63, tab. 52; Fl. Intramong. 4, tab. 45, fig. 1–6.

Described from South. Europe. Type in London (Linn.).

Along banks of toirims, in solonchaks, lower zones of ravines, along banks of rivers and lakes, along irrigation ditches and along fringes of irrigated fields.

IA. Mongolia: *Alash. Gobi* (nor., near Chirgu-bulak spring, July 29, 1873 – Przew.; "Chungwei, on edges of cultivated fields, No. 235, May 1923, R.C. Ching" – Walker, l.c.; "Alashan cent. and Luntou-shan'" – Fl. Intramong. l.c.), *Ordos* (ad ripam sinistram fl. Hoangho infra opp. Hekou, Aug. 5, 1884 – Pot.; "Dalatchi, Otokachi" – Fl. Intramong. l.c.), *Khesi* (valley of Sulekhe river west of Yuimyn' near Padaogo village on road to An'si, Aug. 9, 1875 – Pias.; ad fl. Heiho supra opp. Gaotai solo arenoso, June 20, 1886 — Pot.; Sachzhou oasis, along scrubs on irrigation ditches, Aug. 1, 1895 – Rob.; "Gobi, Lehmzone u. Sandhugel bei Schuang-tsing-yu sudlich von Sutschou, Aug. 1898, Futterer" – Diels, l.c.; Bouloungir, alt. 1500 m, terrains cultives, June 13, 1908, Vaillant — Danguy, l.c.; valley of Sulekhe river" – Chen et Chou, l.c.).

IB. Kashgar: *Nor.* (Kara-teke, south. slope, along brooks, June 9, 1889 – Rob.; bei Maralbaschi und westwarts gegen Kaschgar, Oct. 1902; auf dem Wege von Kutschii zum Dschanart auf Schutthugeln, May 9, 1903 – Merzb.; Zamuschtagh, terraines, humides, July 30, 1907, Vaillant – Danguy, l.c.), *West.* (Yarkend-dar'ya, 900 m, June 21, 1889 – Rob.; oasis near Kashgar town, along irrigation ditches, July 12, 1929 – Pop.), *South.* (oasis Nija, 1260-1350 m, May 21, 1885 – Przew.; Khotan, June 18, 1890 – Grombch.), *East.* (in valle deserto Fl. Tarim inferior, 660-750 m, Oct.-Dec. 1876 – Przew.).

IC. Qaidam: *Plains* ("Linkhu, Makhai, Siligou, Delinkho, Shalyukhe, Syan'zhida, Alaer, etc. – Drev. rast. Tsinkhaya [Woody Plants of Qinghai] l.c.).

IIA. Junggar: *Jung. Alat.* (Dzhair mountain range, Tuz-agny ravine, oasis, in solonchak, June 10, 1951 – Mois.), *Tien Shan* (ad fontes fl. Ili secus fl. Tekess. inter frutices, June 26, 1877 – Przew.; ad fl. Kasch, 900 m, Sept. 6; ad fl. Kasch. 600-1200 m, Sept. 15–1878; Kasch zwischen Ulutai und Nilki, 900-1200 m, June 30; fl. Algoi im NW von Turfan, 1500-1800 m, Sept. 13-1879, A. Reg.; Boro-khoro mountain range between Tol'o and Dzhin-kho, 1889 – Gr.-Grzh.; 18–20 km above Manas town along Manas river, terrace above meadow, June 6; Ili valley 42 km east of bridge on Kash river on road to Ziekty, left bank of Ili-Kunges valley, barren gorge, Aug. 29–1957, Yun., Lee et Yuan'), *Jung. Gobi* (im Thale des Flch. Urungu, June 10-22, 1876 – Pewz.; Adak, June 17, 1877 – Pot.; Takiansi, untere Borotala, Aug. 24, 1877 – A. Reg.; "Ebi-Nor, July 29, 1895, Chaff." – Danguy, l.c.; Mukurtui area west of Ulyun-gur lake, takyr (clay-surfaced desert), June 21, 1908 – Sap.; zwischen Schicho und Dschincho, Sept. 23-25, 1908 – Merzb.; left bank of Manas river 54 km nor. – nor.-east of Podai state farm on road to Chugai settlement, solonchak in sand depressions between ridges, June 17; same site in

104

Chugai settlement, poplar tugai (vegetation-covered bottomland) on second terrace, June 18–1957, Yun., Lee et Yuan'; lower Bulgan-gol 10 km below Bulgan somon, in gorge, June 7, 1964—Davazamc, Danert et Hanelt; ca. 15 km ostlich Jarantai Station, trockene Solontschak der Flussniederung, June 1964—Davazamc, Danert et Hanelt), Zaisan (left bank of Chern. Irtysh in Mai-kain area, hummocky sand, June 7, 1914—Schischk.), *Dzhark.* (Chojur-Sumun, linkes Iliufer, May 26; 1-es Sumin sudwest von Kuldscha, May 29; iliufer bei Kuldscha, May 30; Suidun, westl. von Kuldscha, July 16 —1877, A. Reg.).

General distribution: Aralo-Casp., Fore Ballkh., Nor. and Cent. Tien Shan; Europe (south.), Mediterr., Balk.-Asia Minor, Fore Asia, Caucasus, Mid. Asia (plains), West. Sib. (Irt., Altay), East. Sib. (Ang.-Sayan., Daur), Far East, China (Dunbei, Nor., Nor.-West., Cent., East.).

2. A. pictum Schrenk in Bull. phys.-math. Ac. St.-Petersb. 2 (1844) 115; Ledeb. Fl. Ross. 3 (1847) 43; Fl. Kirgiz. 8 (1959) 204; Fl. Kazakhst. 7 (1964) 125; Grub. Opred. rast. Mong. [Key to Plants of Mongolia] (1982) 204. — *A. hendersonii* Hook. f. in Henderson et Hume, Lahore to Jarkand (1873) 327; Beguinot et Belosersky, Revis. monogr. Apocynum (1913) 78; Persson in Bot. notiser (1938) 298; Chen et Chou, Rast. Pokrov dol. r. Sulekhe [Vegetational Cover of Sulekhe River Valley] (1937) 83, 89; Fl. Kirgiz. 8 (1959) 205; Fl. Kazakhst. 7 (1964) 125; Grub. Opred. rast. Mong. [Key to Plants of Mongolia] (1982) 204. — *A. grandiflorum* Danguy in Bull. Mus. nat. hist. natur. 17 (1911) 340 and 20 (1914) 74. — *Poacynum pictum* (Schrenk) Baill. in Bull. Soc. Linn. Paris, 1 (1888) 757; Woodson in Ann. Missouri Bot. Gard. 17 (1930) 166; Pobed. in Fl. SSSR, 18 (1952) 660; Fl. Sin. 63 (1977) 161; Butkov in Opred rast. Sr. Azii [Key to Plants of Mid. Asia] 8 (1986) 58; [Drev. rast. Tsinkhaya–Woody Plants of Qinghai] (1987) 555; Fl. Intramong. 4 (1993) 114. — *P. hendersonii* (Hook. f.) Woodson in Ann. Missouri Bot. Gard. 17 (1930) 67; Pobed. in Fl. SSSR, 18 (1952) 661; Grub. Konsp. fl. MNR [Conspectus of Flora of Mongolian People's Republic] (1955) 227; id. in Bot. mat. (Leningrad) 19 (1959) 548; Fl. Sin. 63 (1977) 163; Butkov in Opred. rast. Sr. Azii [Key to Plants of Mid. Asia] 8 (1986) 52; [Drevesn. rast. Tsinkhaya–Woody Plants of Qinghai] (1987) 553.

—Ic.: Fl. SSSR, 18, Plate 35, fig. 3 and 5; Fl. Sin. 63, tab. 53 and 54; Fl. Intramong. 4 (1993) tab. 45; Grub. Opred. rast. Mong. [Key to Plants of Mongolia] Plate 109, fig. 490.

Described from East. Kazakhstan (Chu river valley). Type in St.-Petersburg (LE).

In puffed, moist and crusty solonchaks, chee grass thickets, along banks of saline lakes and springs, solonetz-like sand, tugais and along banks of brooks and irrigation ditches.

IA. Mogolia: *West. Gobi* (34 km west of Altay somon centre [Bayan-obo bay] on road to Khairkhan-somon bay, hummocky lowland, puffed hummocky solonchak, Aug. 20, 1979—Grub., Dariima, Muld.; "Bilgekhu-bulak" —Grub. l.c.), *Alash Gobi* (ad. fl. Hei-ho,

July 8; Edzin river between Mumin village and Tufyn, July 11; left bank of Edzin river opposite Mumin town, July 19; between Gantsy-dzak area and Koko-buryuk, July 20; left bank of Edzin river below Shu-bugur area, Aug. 8–1886, Pot.; Alashan, Khamata area, May 19; road from Alashan to Urgu, Shara-burdu area, June 20–1909, Czet.; desert east of valley of Edzin-gol midcourse some 7 versts-1 verst = 1.067 km—south-east of Khara-Khoto town, on sand among mounds with tamarisk, June 7; Edzin-gol valley near upper Ontsin-gol, Bukhan-khub area, on sand, June 10–1926, Glagolev; "Luntou-shan'" —Fl. Intramong. l.c.), *Khesi* (desert south-east of Koutai-syan' [Goutai] town, sand, July 28, 1875—Pias.; Sachzhou oasis [Bulyungir river] June 11, 1879—Przew.; inter oppida Gaotai et Tuiting, June 13; inter vicas Jangta-dsa et Huang-tsheng in salsis, June 23; ad fl. Heiho infra opp. Gaotai, June 25; prope Taloulin, July 5–1886, Pot.; [Satschou] 1890 —Martin; Sachzhou oasis, 1109 m, along irrigation ditches, Aug. 1, 1895—Rob.; "Oasis de Cha-tcheou, Kotatsing-tse, alt. 1000 m, June 8, 1908; Vaillant"—Danguy, l.c.; "valley of midcourse of Sulekhe river"—Chen et Chou, l.c.).

IB. Kashgar: *Nor.* (Kara-Teke, south. slope, 1500–2100 m near brook, June 9 and June 10–1889, Rob.; Chakar, June 9, 1890—Grombch.; zwischen Dscham und Ar-su, Mitte May 1903; bei Outatur, June 19–21, 1903; zwischen Tschartschi und Ueschma in der Steppe, June 3–5, 1908 - Merzb.; "Cha-Jar, May 1907; Tchark-Tchi, route de Bougour a Kourla, Sept. 9, 1907, Vaillant"—Danguy, l.c.; Maralbashinsk oasis, Charbakh village, desert, April 26, 1909—Divn.; "Akso, 1030 m, July 2, 1925"—Persson, l.c.; Maral-bashi oasis, in irrgation ditches, Aug. 3; between Aksu and Kucha near Dzhurga picket, Aug. 13–1929, Pop.), *West.* ("within 10 miles of Jarkand, 1870"—Henderson, l.c.; in Kashgar mountains, June 1872—Kaul' bars; 6 versts - 1 verst = 1.067 km—from Syuget kishlak— village in Central Asia—July 21, 1913—Knorring; "Kashgar, 1925; Okomazar, 1100 m, June 20, 1932"—Persson, l.c.; upper course of Kizyl-su river above Kashgar near Shur-bulak village, July 4; Kashgar oasis, near Faizabad town, July 31; between Kashgar and Maralbash, near Ordenlyk village, in Kizyl-su floodplain, Aug. 1–1929, Pop.; along Charlung brook, 2800 m, Aug. 1, 1941—Serp.), *South.* (Nia oasis, in paludibus frequens et gregarium, April 2 [June 2] 1885—Przew.; "Karauhulik-kol, fresh-water lake at the right side of Lower Tarim, 830 m, May 20, 1900"—Hedin, l.c.; valley of Tarim river 15–20 km below Yarkend- and Khotan-dar'i villages, Sharchakly area, chingil—rock glacier —tugai, Sept. 24, 1957—Yun., Lee et Yuan'), *East.* (Hami oasis, in argilla salsa, June 2, 1879—Przew.).

IC. Qaidam: Plains ("Qaidam"—Drev. rast. Tsinkhaya—Wooded Vegetation of Qinghai, l.c.).

IIA. Junggar: *Jung. Alat.* (Dzhair mountain range, Karamai, solonetz, Aug. 1, 1951— Mois.), *Jung. Gobi* (Umgegend van Gutschen, July 1876—Pewz.; "Bords de l'Ebi-nor, June 29, 1895, Chaff."—Danguy, l.c.; along south, bank of Ebi-nor lake, marshy hummocky sand, Oct. 16, 1929—Pop.; Uinchi somon, Ganshun-usu area, solonchak, lowland, July 31, 1947—Yun.; 4-5 km nor. of Shikho town on road to Chipeitsza state farm, June 28; south-east. vicinity of Ebi-nor lake 6 km nor. of Kum-bulak, near salt plant, solonchak, July 11–1957, Yun., Lee et Yuan'; 40 km south-west of Bulgan somon, on solonchak, Aug. 6; 50 km south-west of Bulgan somon, chee grass thicket, Augt. 7– 1977, Volk. et Rachk.).

General distribution: Aralo-Casp., Fore Balkh.; Mid. Asia (Alay foothills).

Note. *Apocynum hendersonii* Hook. f. is only a large-leaved, large-flowered variety of common dogbane found throughout its distribution range but predominant in the southern parts of the range; it is intimately associated and invariably inseparable from small-flowered, small-leaved intermediate forms.

## Family 99. ASCLEPIADACEAE R. Br.

1.  Pollen grains glued into 4 pellets; stamen spoon-shaped; lobes of corona aristate. Flowers rather few, 2–4 each in leaf axils. Leaves lanceolate with long tapered cusp ............................. ............... ....................................................... 1. Periploca L. (*P. sepium* Bge.).

+   Pollen grains aggregated into 2 pellets (pollen mass), suspended on fine strands to a red corpuscle; lobes of corona without arista, scarious. Flowers many in racemes or cymes ....... ................................................................. 2. Cynanchum L.

### 1. Periploca L.

Gen. pl. ed. 5 (1754) 100

1. P. sepium Bge. Enum. pl. China bor. (1832) 43; Decne. in DC. Prodr. 8 (1844) 198; Franch. Pl. David. 1 (1884) 207; Forbes et Hemsl. Index Fl. Sin. 2 (1889) 101; Kom. Fl. Man'chzh. 3 (1905) 282; Pobed. in Fl. SSSR, 18 (1952) 666; Fl. Sin. 63 (1977) 233; Fl. Intramong. 4 (1993) 116.

—Ic.: Fl. Sin. 63, tab. 94, fig. 1–6; Fl. Intramong. 4, tab. 47.

Described from Nor. China (vicinity of Beijing). Type in Paris (P).

On loessial mounds, consolidated and growing sand dunes.

IA. Mongolia: *Alash. Gobi* (mont. Alaschan australis, in arenoso-limosis frequens, June 11 [23] 1872; ibid, Aug. 20, 1880—Przew.; Alashan, Tsokto-khure temple, Ikhe-khuduk well, 1350 m, Sept. 25, 1901—Lad.), *Ordos* (in collibus arenae mobilis a lacu [fl. !] Narin-gol S versus, Sept. 10 and 11, 1884—Pot.; "south. regions of Ulantsab and Ordos, south-east. Alashan"—Fl. Intramong. l.c.).

General distribution: Far East (Ussur.), China (Dunbei, Nor., Nor.-West., Cent., East., South-West.).

### 2. Cynanchum L.

Sp. pl. (1753) 212; id. Gen. pl., ed. 5 (1754) 101.— *Vincetoxicum* N.M. Wolf, Gen. Pl. (1776) 130.—*Seutera* Reichb. Consp. regni veg. (1838) 131. — *Cynoctonum* E. Mey. Comm. pl. Afr. Austr. (1837) 215, non Gmel. 1791. —*Pycnostelma* Bge. ex Decne in DC. Prodr. 8 (1844) 512.—*Antitoxicum* Pobed. in Fl. URSS, 18 (1952) 752.

1.  Lianas with long and slender climbing stem ............................ 2.

+   Plants with erect, simple or branched stem (or stems), 20–100 cm tall ............................................................................... 6.

2.  Leaves hastate or cordate with deeply emarginated or (rarely) truncated base ............................................................................... 3.

+   Leaves lanceolate, with orbicular or broad-deltoid base ............. ................................................................. 4. C. gobicum Grub.

3. Leaves hastate, with narrow-deltoid cuspidate blade and orbicular basal lobes. Plants glabrous or dispersely pilose ....... 4.

+ Leaves orbicular-cordate, short-cuspidate. Entire plant compactly pubescent with short flexuose hairs ............................
...................................................................... 2. C. chinense R. Br.

4. Basal lobes of leaf blade set downward and laterally and notch between them open, broad ............................................................ 5.

+ Basal lobes of leaf blade set downward and inward such that they overlie one another and notch between them narrow and closed ................................................................ 3. C. heydei Hook. f.

5. Lateral lobes of leaf blade wide-spaced and notch between them truncated at tip, trapezoid; sometimes, blade and lobes elongated, narrow, and leaf sagittate. Root tubercular. Flowers greenish yellow or white ................................. 1. C. bungei Decne.

+ Lateral lobes of leaf blade proximated and notch between them deltoid. Flowers from pink to purplish red. Root rachiform ........
................................................................ 5. C. sibiricum Willd.

6. Leaves broad, from broad-ovoid to lanceolate-ovoid ................ 7.

+ Leaves from linear to narrow-lanceolate ................................... 10.

7. Leaves sessile, broad-ovoid or suborbicular, 0.8–1.8 cm long, 0.7–1.6 cm broad, truncated or weakly cordate at base with short-cuspidate or obtuse tip. Stems several, 15–30 cm tall, emerging from multicipital ligneous rhizome ...............................
............................................................ 11. C. pusillum Grub.

⊢ Leaves on distinctly manifest, although short, petioles, very large, 4–10 cm long. Root fibrous and stems single or more, 40–80 cm tall ................................................................................ 8.

8. Leaves obovoid, with short auricles at base, semiamplexicaul, short-cuspidate ......... 6. C. amplexicaule (Sieb. et Zucc.) Hemsl.

+ Leaves with orbicular or cuneate base ........................................ 9.

9. Leaves oval, sharply narrowed into short cusp; on both sides, like stem, with compact white hairs ............. 7. C. atratum Bge.

+ Leaves lanceolate-elliptical or broad-lanceolate, gradually narrowed into long cusp, puberulent only underneath along nerves and margin ........................................................................
.................................... 8. C. hanckokianum (Maxim.) Al. Iljinoki.

10. Leaves narrow-lanceolate, 7–10 (15) cm long, 5–10 (15) mm broad. Root fibrous. Stem single, rarely 2–3, 30–60 cm tall ..... 11.

+ Leaves linear or linear-lanceolate, 3–7 cm long, 2–5 mm broad. Stems few, emerging from multicipital ligneous rhizome, 15–30 cm tall ................................................................................. 12.

11.  Stem simple, branched only in inflorescence. Latter terminal, paniculate; flowers greenish yellow ...................................................
.................................................... 10. C. paniculatum (Bge.) Kitag.

+  Stem branched from lower half or simple. Cymes axillary all along stem; flowers violet-red ......... 9. C. komarovii Al. Iljinski.

12.  Stem somewhat flexuose, branched in upper half and dense-foliate. Inflorescence fascicular at tip of stems and branches; flowers greenish yellow, 3–3.5 mm in diam. ...........................
.................................................... 13. C. thesioides (Freyn) K. Schum.

+  Stems erect, dichotomously branched in upper part, somewhat foliate. Cymes at tip of stems and branches; flowers purple-red, large, 10–15 mm in diam. Entire plant diffusely hispid .............
.................................................... 12. C. purpureum (Pall.) K. Schum.

1. **C. bungei** Decne. in DC. Prodr. 8 (1844) 549; Franch. Pl. David. 1 (1884) 209; Forbes et Hemsl. Index Fl. Sin. 2 (1889) 105; Kom. Fl. Man'chzh. 3 (1905) 295; Danguy in Bull. Mus. nat. hist. natur. 17 (1911) 340; Kitag. Lin. Fl. Mansh. (1939) 362; Chen et Chou, Rast. pokrov dol. r. Sulekhe [Vegetational Cover of Sulekhe River Valley] (1957) 89; Fl. Sin. 63 (1977) 322; Fl. Xizang, 4 (1985) 14; Pl. vasc. Helanshan (1986) 205; Fl. Intramong. 4 (1993) 130. — *Asclepias hastata* Bge. Enum. pl. China bor. (1832) 48.

—Ic.: Fl. Sin. 63, tab. 114, fig. 8–13; Fl. Xizang, 4, tab. 7, fig. 15–18; Fl. Intramong. 4. tab. 53. fig. 10–14.

Described from Nor. China (vicinity of Beijing). Type in Paris (P). Isotype in St.-Petersburg (LE).

On steppe slopes of mounds and mountains, along precipices and banks of rivers among shrubs, on growing sand dunes (barhans).

IA. Mongolia: *East. Mong.* ("Baotou" —Fl. Intramong. l.c.), *Alash. Gobi* ("Alashan" Fl. Intramong. l.c.; "Helanshan" [Alashan mountain range]—Pl. vasc. Helanshan, l.c.), *Khesi* ("valley of Sulekhe river" —Chen et Chou, l.c.).

IIIA. Qinghai: *Nanshan* ("Environs de Sining-Fou, alt. 2400 m, July 18, 1908, Vaillant" —Danguy, l.c.).

General distribution: China (Dunbei, Nor., Nor.-West., Cent., East.).

2. **C. chinense** R. Br. in Mem. Wern. Soc. 1 (1810) 44; Decne. in DC. Prodr. 8 (1844) 548; Forbes et Hemsl. Index Fl. Sin. 2 (1889) 105; Kom. Fl. Man'chzh. 3 (1905) 288; Kitag. Lin. Fl. Mansh. (1939) 362; Walker in Contribs U.S. Nat. Herb. 28 (1941) 653; Fl. Sin. 63 (1977) 314; Pl. vasc. Helanshan (1986) 205; Fl. Intramong. 4 (1993) 128. — *C. pubescens* Bge. Enum. pl. China bor. (1832) 44; Danguy in Bull. Mus. nat. hist. natur. 17 (1911) 360; Grub. Konsp. fl. MNR [Conspectus of Flora of Mongolian People's Republic] (1955) 227; Chen et Chou, Rast. pokrov dol. r. Sulekhe

[Vegetational Cover of Sulekhe River Valley] (1957) 89; Grub. Opred. rast. Mong. [Key to Plants of Mongolia] (1982) 204.

—Ic: Fl. Sin. 63, tab. 109, fig. 1–9; Grub. Opred. rast. Mong. [Key to Plants of Mongolia] Plate 109, fig. 493; Fl. Intramong. 4, tab. 53, fig. 1–9.

Described from Nor. China (vicinity of Beijing). Type in London (BM ?).

On fixed sand dunes and hummocky as well as thin solonetzic sand, in chee grass thickets, along flanks and floors of large gorges, among shrubs on slopes of mountains and knolls.

IA. Mongolia: *East. Mong.* (Muni-ula, decliv. australis, solo limoso-arenoso, July 21, 1871 – Przew.; planities circa Kuku-hoton, urbis Futscheu, Aug. 2; in collibus arenae mobilis ripa dextrae fl. Hoangho infra Hekou, Aug. 7–1884, Pot.; "Datsinshan' " – Fl. Intramong. l.c.), *East. Gobi* (East. Gobi, floor of gully, on fine rubble, Aug. 16, 1926 – Lisovsky; south of Galba mountain range, Dzun-khuris area (Bayan-Mogo-gol], nitrous mounds, April 23, 1940 – Yun.; on Khubsugul somon to Sain-Shanda town road, on floor of large ravine, Aug. 3, 1970 – Grub., Ulzij., Tserenbalzhid), *Alash. Gobi* (Tengeri-elisyn sand, Sept. 1901 – Lad.; Tengeri sand, Shangyn-dalai area, on sand dune (barhans), July 8, 1908 – Czet.; "Chung-wei, along roadsides, No. 221, June 1923, Ching" – Walker, l.c.; Arshantyn-nuru mountain range 8 km west of Shuulin post, near spring, along ravine, Aug. 1; Bordzon-Gobi south-west of Undur-Bogdo-ula along large ravine in red formations, Aug. 2; south-east of same mountain along border road, in large ravine with elms, Aug. 3, 1989 – Grub., Gubanov, Dariima; "Helanshan" [Alashan mountain range] – Pl. vasc. Helanshan, l.c.), *Ordos* (Ordos, 1884 – Bretschneider; valle fl. Ulan-Morin, Aug. 22, 1884 – Pot.), *Khesi* ("Champs Kantcheou, Kansu, June 29, 1908, Vaillant" – Danguy, l.c.; "valley of Sulekhe river" – Chen et Chou, l.c.).

IIIA. Qinghai: *Amdo* (prope Huidui ad fl. Hoangho, June 13, 1880 – Przew.).

General distribution: China (south. Dunbei, Nor., Nor.-West., Cent., East., South).

109 **3. C. heydei Hook. f. Fl. Brit. India, 4 (1889) 25; Fl. Sin. 63 (1977) 313; Fl. Xizang, 4 (1985) 14.**

—Ic.: Fl. Xizang, 4, tab. 7, fig. 10.

Described from Himalayas (Kashmir). Type in London (K).

On forest-covered mountain slopes.

IIIB. Tibet: *Chang Tang* ? ("West. Tibet" – Fl. Xizang, l.c.).

General distribution: Himalayas (Kashmir).

Note. Reliable finds of this species in Tibet have not been cited either in Fl. Sin. or Fl. Xizang; evidently, Chinese botanists did not possess its specimens nor did they cite the data source for including *C. heydei* in the floral composition of Tibet.

**4. C. gobicum Grub. in Nov. syst. pl. vasc. 32 (2000) 135.** — *Antitoxicum lanceolatum* Grub. in Not. syst. (Leningrad) 17 (1955) 21, non *Cynanchum lanceolatum* Poir. 1811; Grub. Konsp. fl. MNR [Conspectus of Flora of Mongolian People's Republic] (1955) 227. — *Vincetoxicum lanceolatum* (Grub.). Grub. in Nov. syst. pl. vasc. 21 (1984) 208; Grub. Opred. rast. Mong. [Key to Plants of Mongolia] (1982) 204, 416; Gubanov, Konsp. fl. Vneshn. Mong. [Conspectus of Flora of Outer Mongolia] (1996) 85.

—Ic.: Grub. l.c. (1982), Plate 109, fig. 494.

Described from Mongolia (Gobi Alt.). Type in St.-Petersburg (LE). Map 4.

On rocks and rocky slopes and trails, along gorges and ravines.

IA. Mongolia: *Gobi Alt.* (Dzun-Saikhan mountain range, commencement of northern trail along road to pass from Dalan-Dzadagad, on rocks in gorge, July 22, 1943, Yun.— typus !; Khuren-Khana mountain range, beginning of wide Musarin-khundii gorge, about 1800 m, on rocks, Sept. 7, 1979—Grub., Dariima, Muld; south. foothills of west. part of Gurban-Saikhan mountain range, Bayan-Undur mountains, on rocky slopes, Aug. 6, 1981—Gub. [MW]), *East. Gobi* (Saikhan-Dulan somon, 20 km south-west of Sain-shanda town, hammada—rocky desert—July 8, 1975—Zhurba [MW]; 15 km south of Dzun-Bayan town, Takhyat-ula mountain, 500 m above sea level, along floor of gorge, July 7; Khan-Bogdo-ula mountains, 20 km east—south-east of Khan-Bogdo settlement, 1000 m above sea level, along ravines, Sept. 3, 1982—Gub. [MW]), *Alash. Gobi* (8 km east of Shuulin border post, in intermittently flooded sites, in ravines, Aug. 4, 1981— Gub. [MW]; 8 km west of Shuulin border post, Arshantin-nuru mountain range, on slopes, Aug. 1, 1989—Grub., Gub., Dariima).

General distribution: endemic.

Note. The reference of Gubanov (l.c.). to the find of this species in Western and Junggar Gobi is not supported by factual data.

5. C. sibiricum Willd. in Ges. Naturf. Fr. Neue Schr. (1799) 124; Kom. Fl. Man'chzh. 3 (1905) 291; Pobed. in Fl. SSSR, 18 (1952) 716; Grub. Konsp. fl. MNR [Conspectus of Flora of Mongolian People's Republic] (1955) 227; Fl. Sin. 63 (1977) 311; Grub. Opred. rast. Mong. [Key to Plants of Mongolia] (1982) 204; Fl. Xizang, 4 (1985) 14; R. Vinogradova in Opred. rast. Sr. Azii [Key to Plants of Mid. Asia] 8 (1986) 61. — *C. acutum* auct. non L.: Ledeb. Fl. alt. 1 (1829) 278; Hedin, S. Tibet, 6, 3 (1922) 47; Persson in Bot. notiser (1938) 298; Kryl. Fl. Zap. Sib. 9 (1937) 2208; Fl. Kirgiz. 8 (1959) 207; Fl. Kazakhst. 7 (1964) 129. — *C. acutum* L. β *longifolium* Ledeb. Fl. Ross. 3 (1847) 548; Boiss. Fl. or. 4 (1875) 60; Danguy in Bull. Mus. nat. hist. natur. 17 (1911) 340.— *C. cathayense* Tsiang et Zhang in Acta phytotax. sin. 12 (1974) 110; Fl. Sin. 63 (1977) 379; Fl. Intramong. 4 (1993) 126.

—Ic.: Willd. l.c. tab. 5, fig. 2; Fl. Sin. 63, tab. 107 and 137; Fl. Intramong. 4 (1993) tab. 51 (sub nom. *C. cathayense*).

Described from Siberia. Type in Berlin (B)?

110 On growing sand dunes (barhans) and hummocky sand, on sandy-pebbly ravine floors, solonetzic chee grass thickets, sand-covered trails of mountains and knolls, tugais, among shrubs around springs and brooks, fringes of fields and in oasis in desert zone.

IA. Mongolia: *Depr. Lakes* (inter finis Sibiriae et Chobdo 1870—Kalning; Airag-nur lake, east coast, rubbly-sandy coastal strip, Aug. 12, 1972—Metel'tseva), *Val. Lakes* ("Tatsiin-gol" —Grub. l.c.), *Gobi Alt.* (in latere australi jugi Bain-tsagan, Aug. 22 and 23, 1886—Pot.; Bayan-Tsagan low mountain range near Shirigiin-khuduk well, Sept. 3, 1927

—Simukova; west. end of Dzolen-Bogdo mountain range, trail, Aug. 4, 1947—Yun.; hummocky area 25 km nor. of Nemegetu mountain range around Barun-boro-obo mountains and Khara-tologoi well, Aug. 3, 1948—Grub.), *East. Gobi* (5 km nor.-east of Zhaal-Shanda well on road to Dalan-Dzadagad town, July 27, 1972—Guricheva et Rachk.; "Ulan-Khoto" —Fl. Intramong. l.c.), *West. Gobi* (Bayan-Gobi somon, Shara-khulusun oasis, Aug. 9, 1943—Tsebigmid; Bayan-Undur somon, Burkhantu-bulak spring, Aug. 23, 1948—Grub.; 15 km south-east of Alag-Shanda well, on sand among saxaul, June 28, 1974—Rachk. et Volk.; 100 km west—south-west of Eikhin-gol oasis, in tamarisk shrubs, Aug. 26, 1976—Rachk. et Damba; knolls around Khutsyn-Shanda spring, sand knolls in gorge, Sept. 2, 1979—Grub., Dariima, Muld.), *Alash. Gobi* (Alashan boundary, Burgasne-amne-usu spring, Aug. 29, 1924—Pakhom; 5 km south of Obtu-khural [post] on road to Sogo-nuru, rocky desert, July 27, 1943—Yun.; Bordzon-Gobi, hummocky area nor.-east of Khaldzan-ula, on rock in ravine, Sept. 8, 1950—Lavr., Kal., Yun), *Khesi* (Dan-khe river, Satsza-yuan'-tszy, 1200 m, along shrubs, July 28, 1895 —Rob.; "Cha-tceou [Sa-chzhou], Touen-houang [Dunkhuan], June 4, 1908, Vaillant" —Danguy, l.c.).

IB. Kashgar: *Nor.* (Kara-Teke, 1500–2100 m, in shrubs, June 9, 1889; Khaidyk-gol river, Chubogorin-nor area, 1200 m, ploughed land, Aug. 16, 1893—Rob.; Uch-Turfan, June 9, 1908—Divn.; near Taz-Lyangar village, Aug. 11; near Yangi-Abad village, Aug. 19, 1929—Pop.; "Province de Koutchar, June 1907, Vaillant" —Danguy, l.c.; "Maralbaschi, 1120 m, June 17, 1932" —Persson, l.c.), *West.* (Tir village, Aug. 11, 1890—Grombch.; Bakh village on Charlym river, Aug. 6, 1909—Divn.; Kashgar oasis around Yangi-gisar, along fields, July 25, 1929—Pop.; Mia [Tkesekrik] river, 2300–2800 m, July 16, 1941—Serp.; "Jarkend, 1350 m, 1931; Kashgar, 1130 m, July 16, 1934" —Persson, l.c.), *South.* (Kcria, Jassulgun oasis, May 27 [June 8]; Keria, in pago Awak ad fl. Ashi-Daria, 1600 m, Aug. 1–1885, Przew.; Khotan, June 18, 1890—Grombch.; nor. slope of Russky mountain range, Mal'dzha river, 2450 m, July 17, 1890—Rob.; "Karaumelik-köl, freshwater lake on the right side of Lower Tarim, 880 m, May 20; Lower Tarim about 830 m, early summer of 1900; Tunataghdi, Lower Tarim, 825 m, June 8–1900" —Hedin, l.c.), *East.* (Gobi [Ergou brook], Aug. 22, 1875—Pias.; fl. Algoi in NW von Turfan, 1500–1800 m, Sept. 13, 1879—A. Reg.; Chitkal' area Sept. 14, 1898—Klem.; "Ach-Ma, alt. 1100 m, pres de Bougour, Sept. 8, 1907, Vaillant" —Danguy, l.c.).

IIA. Junggar: *Cis-Alt.* (Altai australis [? vall. fl. Kansagatai], Sept. 1876—Pot.), *Jung. Alat.* (Dzhair mountain range, Tuzagny ravine, solonchak, June 27, 1951—Mois.), *Tien Shan* (Dschagastai, sudlich von Kuldscha, 1500–2100 m, Aug. 9, 1877—A. Reg.; ad Ili fl. super., in ripa fl. Tekess, July 8, 1877—Przew.; Arustan ad fl. Ili, Aug. 7, 1878—Fet.; Ili-Kunges valley 42 km east of bridge on Kash river on road to Ziekta, right bank, at bottom of lateral gorge, Aug. 29, 1957—Yun., Lee, Yuan'), *Jung. Gobi* (Kharm, Sept. 30, 1875—Pias.; untere Borotala, Aug. 1878—A. Reg.; in litore abrupto austr. lacus Uljungur, Aug. 12, 1876; ad fl. Nom. limosis, June 17, 1877—Pot.; nor. of Guchen, Gashun, 1889 —Gr.-Grzh.; between Santa-khu and Dzhimuchi, in wheat field, Aug. 19; vicinity of Fukan, Aug. 21–1898, Klem; nor. bank of Ulyungur lake, barren slopes, June 23, 1908—Sap.; 14 km from Bodonchiin-baishing on Guchen road, upper terrace of Bodonchiin-gol river, on sand, Sept. 12; south. foothills of Barangiin-Khara-nuru mountain range, gorge 5–8 km west of Mergen-ula, Sept. 19–1948, Grub.; valley of Derbaty river at its intersection with Karamai—Altai [Shara-sume] highway, floodplain, solonchak, reed grass thicket, June 20, 1957—Yun., Lee, Yuan'; 2 km nor.-east of Gashun-us spring, on solonchak, Aug. 10, 1977—Rachk. et Volk.), Dzhark. (ad fl. Laba, 300–420 m, 1874— Larionoff; Kuldscha, 1876—Golicke; Chojur-Sumun sudlich von Kuldscha, 540–600 m, May 26, ad fl. Ili, May 27; S-tes Sumun [Iliufer], May 28, 29; Pilutschi prope Kuldscha, June 17; Iliufer westl. von Kuldscha, June; Suidum, July 16–1877; Suidua, July 1878—A. Reg.).

111

128

General distribution: Aralo-Casp., Fore Balkh., Jung.-Tarb.; Mid. Asia (plain and mount. Turkm.), West. Sib. (south. Altay and south. Irtysh region), China (Nor.: Hebei).

Note. 1. *C. cathayense* Tsiang et Zhang, l.c., is undoubtedly a synonym of *C. sibiricum* Willd. According to the authors, the distinctive features of this species are the red colour of flowers and absence of inner liguliform appendages in lobes of corona and they even place it in a different section *Cyathella*. These two characteristics, however, are applicable to *C. sibiricum* Willd. as well; it differs from the closely related western *C. acutum* L. in the red colour (from pink to dark purple) of flowers and partial reduction of liguliform appendages in corona right up to their total absence. This has been pointed out by Pobedimova as well in Flora SSSR (l.c.).

2. One more from this group of herbaceous lianas, East Asian *Cynanchum wilfordii* (Maxim.) Hemsl., is found in Kashgar oasis in Sinkiang but it has been imported by the Chinese and cultivated as a medicinal and edible plant.

6. **C. amplexicaule** (Sieb. et Zucc.) Hemsl. in J. Linn. Soc. London (Bot.) 26 (1889) 104; Kom. Fl. Man'chzh. 3 (1905) 286; Kitag. Lin. Fl. Mansh. (1939) 362; Fl. Sin. 63 (1977) 332; Fl. Intramong. 4 (1993) 118.— *Vincetoxicum amplexicaule* Sieb. et Zucc. in Abh. Ak. Wiss. München, 4, 3 (1846) 162.— *Antitoxicum amplexicaule* (Sieb. et Zucc.) Pobed. in Fl. URSS, 18 (1952) 705.

—Ic.: Fl. Sin. 63, tab. 116; Fl. Intramong. 4, tab. 48, fig. 1–7.

Described from Japan. Type in Munich (M).

On steppe slopes and meadows, sand shoals and trails.

IA. Mongolia: *East. Mong.* ("Kulun-Buir-nor" —Fl. Intramong. l.c.).

General distribution: Far East (Ussur.), China (Dunbei, Nor., Nor.-West., Cent., East.), Japan.

Note. Variety with dark brown (not greenish yellow) flowers, var. *castanea* Makino, is also found.

7. **C. atratum** Bge. Enum. pl. China bor. (1832) 45; Forbes et Hemsl. Index Fl. Sin. 2 (1889) 104; Kom. Fl. Man'chzh. 3 (1905) 287; Kitag. Lin. Fl. Mansh. (1939) 362; Fl. Sin. 63 (1977) 344; Fl. Intramong. 4 (1993) 120. — *Vincetoxicum atratum* Decne. in DC. Prodr. 8 (1844) 523; Maxim. in Bull. Ac. Sci. Petersb. 23 (1877) 362.— *Antitoxicum atratum* (Bge.) Pobed. in Fl. URSS, 18 (1952) 704.

—Ic.: Fl. SSSR, 18, Plate 38, fig. 4; Fl. Sin. 63, tab. 123; Fl. Intramong. 4, tab. 48, fig. 8–9.

Described from Nor. China. Type in Paris (P). Isotype in St.-Petersburg (LE).

In meadows and meadowy slopes, along ravines.

IA. Mongolia: *East. Mong.* ("Datsinshan', valley of Shilin-gol river" —Fl. Intramong. l.c.).

General distribution: Far East (south.: Zee-Bur., Ussur.), China (entire country).

8. **C. hanckokianum** (Maxim.) Al. Iljinski in Acta Horti Petrop. 34 (1920) 54; Tsiang et P.T. Li in Acta phytotax. sin. 12 (1974) 94; Fl. Sin. 63 (1977) 344; Fl. Intramong. 4 (1993) 122. — *C. mongolicum* Hemsl. in J. Linn.

112   Soc. London (Bot.) 26 (1889) 107; Hao in Bot. Jahrb. 63 (1938) 631; Walker in Contribs U.S. Nat. Herb. 28 (1941) 651. — *Vincetoxicum mongolicum* Maxim. β *hanckokianum* Maxim. in Bull. Ac. Sci. Petersb. 23 (1877) 356. — *V. mongolicum* auct. non Maxim.; Danguy in Bull. Mus. nat. hist. natur. 17 (1911) 340.

—Ic.: Fl. Sin. 63, tab. 122; Fl. Intramong. 4, tab. 49, fig. 12–13.

Described from Nor. China (Syao-utai-shan' mountains). Type in St.-Petersburg (LE).

On plains and hummocky sand and semi-fixed sand dunes (barhans), along bottoms of ravines, sometimes profusely.

IA. Mongolia: *Alash. Gobi* ("Hsincheng", June 1923, Ching—Walker, l.c.; "Alashan" —Fl. Intramong. l.c.), *Ordos* (ad lacum salsum Baga-Tschikyr, Sept. 25, 1884—Pot.; "Ordos"—Fl. Intramong. l.c.), *Khesi* ("Lanchow, an trockenen Standorten um 1700 m, 1930"—Hao, l.c.).

IIIA. Qinghai: *Nanshan* ("Yao-chieh, on bare dry gravelly foothills", June 1923, Ching—Walker, l.c.).

General distribution: China (Nor., Nor.-West., South-West.: Sichuan).

9. **C. komarovii** Al. Iljinski in Acta Horti Petrop. 34 (1820) 52; Tsiang et P.T. Li in Acta phytotax. sin. 12 (1974) 98; Fl. Sin. 63 (1977) 353; Fl. Intramong. 4 (1993) 122; Pl. vasc. Helanshan (1986) 204.— *C. mongolicum* auct. non Hemsl. (1889): K. Schum. in Engler et Prantl, Naturl. Pflanzenfam. 4, 2 (1895) 253; Kom. in Acta Horti Petrop. 34 (1920) 54. — *Vincetoxicum mongolicum* Maxim. in Bull. Ac. Sci. Petersb. 23 (1877) 356.

—Ic.: Fl. Sin. 63, tab. 126; Fl. Intramong. 4, tab. 49, fig. 1–11.

Described from Inner Mongolia (Alash. Gobi). Type in St.-Petersburg (LE).

On sunny rock slopes, solonetzic sand, sandy-pebbly shoals and river banks.

IA. Mongolia: *East. Mong.* ("Shilin-gol"—Fl. Intramong. l.c.), *Alash. Gobi* (Alaschan deserta, [solo] limoso-arenoso frequens, July 12 [24] 1873—Przew., lectotypus !; Tengeri-elisyn sand, Sept. 1901—Lad.; Tengeri sand, Tarbagai area, among sand dunes, July 7, 1908—Czet.), *Ordos* (valle Hoangho limosa, parcius in arenis Kusuptschi, Aug. 2; ibid. prato inundato limoso-arenoso frequens, Aug. 9–1871, Przew., syntypus !; in collibus arenae mobilis circa monast. Bortai-ssume, valle fl. Ulan-Morin, Aug. 21, 1884—Pot.; "Helanshan"—Pl. vasc. Helanshan, l.c.), *Khesi* (south: "Bords du Hoangho, Lantcheou, July 23, 1908, Vaillant"—Danguy, l.c.).

IIIA. Qinghai: *Amdo* (ab ostium fl. Karyn secus fl. Hoangho superiorem, May 5, 1885—Pot.; Xining alps, south. slope of Guiduisha, June 24, 1890—Gr.-Grzh.; e semin. a clar. Czetyrkin in monte Dzhachar anno 1909 collectis in H.B. Petrop. culta, Aug. 7–13, 1919—A. Iljinski).

General distribution: China (Nor.; Hebei).

Note. While re-naming this species, Al. Iljinski (l.c.) cited as type a specimen of plant grown in St.-Petersburg Botanical Garden from seeds collected by S. Czetyrkin in Amdo in 1909 which is not legal. The type should be the herbarium plant specimens cited by K.I. Maximowicz (l.c.) while describing this species for the first time as *Vincetoxicum mongolicum* Maxim.

10. C. paniculatum (Bge.) Kitag. in J. Jap. Bot. 16 (1940) 20; Hara,
Enum. Spermat. Jap. (1948) 153; Fl. Sin. 63 (1977) 351; Fl. Intramong. 4
(1993) 124. — *Asclepias paniculata* Bge. Enum. pl. China bor. (1832) 43
(seors. impr.) — *Pycnostelma chinense* Bge. ex Decne. in DC. Prodr. 8 (1844)
512; Maxim. in Bull. Ac. Sci. Petersb. 23 (1877) 153; Forbes et Hemsl.
113  Index Fl. Sin. 2 (1889) 102; Kom. Fl. Man'chzh. 3 (1905) 281. — *P.*
*paniculatum* K. Schum. in Engler et Prantl, Naturl. Pflanzenfam. 4, 2
(1893) 243; Kitag. Lin. Fl. Mansh. (1939) 364; Pobed. in Fl. SSSR, 18 (1952)
671.— *Vincetoxicum picnostelma* Kitag. in J. Jap. Bot. 16 (1940) 19.

—Ic.: Fl. Sin. 63, tab. 125; Fl. Intramong. tab. 50, fig. 1–5.

Described from Nor. China (nor. of Beijing). Type in Paris (P). Isotype
in St.-Petersburg (LE).

On meadowy steppe slopes, steppe scrubs, floodplain meadows, along
forest borders.

IA. Mongolia: *East. Mong.* ("Kulun-Buir-nor, Datsinshan', Shilingol. [?]" — Fl.
Intramong. l.c.).

General distribution: East. Sib. (Daur.), Far East (south.: Zee-Bur., Ussur.); Nor.
Mong. (Mong.-Daur., Cis-Hing.), China (Dunbei, Nor., Nor.-West., Cent., East., South-
West., South.), Korea, Japan.

11. C. pusillum Grub. in Nov. syst. pl. vasc. 32 (2001) 137.

Described from Sinkiang. Type in St.-Petersburg (LE). Plate I, fig. 4.

IB. Kashgar: *West.* (Kashgar district, between Merket town and Tarim-bazar, in
depressions among sand knolls, saline soil, Sept. 5, 1940 — T. Trofimov, typus !).

General distribution: endemic.

Note. This species is very close to *C. pumilum* (Decne.) Bornm. but differs well from
it in entirely sessile and very small leaves (0.8–1.8 cm long, 0.7–1.6 cm broad, not 3–4
cm long, 2.5–3.5 cm broad, short-petiolate) and leaflets puberulent (not glabrous)
outside. Plants 30–40 cm tall.

12. C. purpureum (Pall.) K. Schum. in Engler et Prantl, Naturl.
Pflanzenfam. 4, 2 (1895) 253; Nakai, Index Fl. Jehol. (1936) 40; Kitag. Lin.
Fl. Mansh. (1938) 363; Fl. Sin. 63 (1977) 373; Fl. Intramong. 4 (1993) 121.
— *Asclepias purpurea* Pall. Reise, 3 (1776) 260.— *Cynanchum roseum* R. Br. in
Mem. Werner. Soc. 1 (1810) 47; Forbes et Hemsl. Index Fl. Sin. 2 (1889)
108; Kom. Fl. Man'chzh. 3 (1905) 290; Danguy in Bull. Mus. nat. hist.
natur. 20 (1914) 15.— *Cynoctonum roseum* Decne. in DC. Prodr. 8 (1844)
532; Franch. Pl. David. 1 (1844) 209. — *C. purpureum* (Pall.) Pobed. in Fl.
URSS, 18 (1952) 709; Grub. Konsp. fl. MNR [Conspectus of Flora of
Mongolian People's Republic] (1955) 227; id. Opred. rast. Mong. [Key to
Plants of Mongolia] (1982) 204.

—Ic.: Fl. SSSR, 18, Plate 39, fig. 2; Fl. Intramong. 4, tab. 50, fig. 6–12.

Described from East. Siberia (vicinity of Irkutsk). Type in London
(BM).

On steppe sandy, rubbly and rocky slopes of knolls and mountains, in arid sandy steppes, on rocks.

IA. Mongolia: *Cent. Khalkha* (Zaan-Shire mountain 80 km east—south-east of Undurkhan town [47°15′ N. lat., 115°35′ E. long.], June 19, 1987—Kam., Gub., Ganbold et al), *East. Mong.* (in monte Batu-Chan inter Kerulen [Urgo] et Dolon-nor, 1870—Lomonossow; Kulun-buir-nor plain, Boro-khoolai, May 5; same site, Mandybai lake, May 31–1899, Pot.; "Environs de Kailar, steppe sablonneuse, June 19, 1896, Chaff."—Danguy, l.c.; 55 km south of Tamtsag settlement, June 17, 1956—Dashnyam; Bayan-Khan-ula mountain on south. extremity of Tareisk highwater area, Aug. 24, 1985—Kam. et Dariima), *Alash. Gobi* ("east. Alashan"—Fl. Intramong. l.c.).

General distribution: East. Sib. (south: Ang.-Sayans, Daur.), Far East (south: Zee-Bur., Ussur.), Nor. Mong. (Hent., Mong.-Daur., Cis-Hing.), China (Nor.: Hebei), Korea (nor.).

13. C. thesioides (Freyn) K. Schum. in Engler et Prantl, Naturl. Pflanzenfam. 4, 2 (1895) 252; Tsing et P.T. Li in Acta phytotax. sin. 12 (1974) 99; Fl. Sin. 63 (1977) 367; Pl. vasc. Helanshan (1986) 204; Fl. Intramong. 4 (1993) 126.— *C. sibiricum* (L.) R. Br. in Mem. Werner. Soc. 1 (1810) 48; Bge. Enum. pl. China bor. (1832) 44; Forbes et Hemsl. Index Fl. Sin. 2 (1889) 108; Palibin in Trud. Troitskosavsko-Kyakhtinsk. otdeleniya Priamursk. otdela Russk. Geogr. Obshch. 7, 3 (1904) 50; Kom. Fl. Sin. Man'chzh. 3 (1905) 291; Hao in Bot. Jahrb. 68 (1938) 631; Kitag. Lin. Fl. Mansh. (1939) 363; Walker in Contribs U.S. Nat. Herb. 28 (1941) 654; Jernakov in Acta pedolog. sin. 2 (1954) 247, non Willd. 1799 — *Vincetoxicum sibiricum* (L.) Decne. in DC. Prodr. 8 (1844) 525; Ledeb. Fl. Ross. 3 (1847) 46; Maxim. in Bull. Ac. Sci. Petersb. 23 (1877) 355; Kryl. Fl. Zap. Sib. 9 (1937) 2207; Fl. Kazakhst. 7 (1964) 128; R. Vinogradova in Opred. rast. Sr. Azii [Key to Plants of Mid. Asia] 8 (1986) 61; Grub. Opred. rast. Mong. [Key to Plants of Mongolia] (1982) 204. —*Asclepias sibirica* L. Sp. pl. (1753) 217. — *Vincetoxicum thesioides* Freyn in Oest. Bot. Zeitschr. 40 (1890) 124. — *Antitoxicum sibiricum* (L.) Pobed. in Fl. URSS, 18 (1952) 707; Grub. Konsp. fl. MNR [Conspectus of Flora of Mongolian People's Republic] (1955) 227.

—Ic.: Fl. SSSR, 18, Plate 38, fig. 1; Fl. Kazakhst. 7, Plate 15, fig. 4; Fl. Sin. 63, tab. 133; Fl. Intramong. 4, tab. 52, fig. 1–11.

Described from Siberia. Type in London (Linn.).

In barren and arid sandy steppes, steppe rubble slopes and talus, along small ravines, on rather thin sand and river shoals, often.

IA. Mongolia: *Khobd., Mong. Alt., Cent. Khalkha, East. Mong., Val. Lakes, Depr. Lakes, Gobi-Alt., East. Gobi, West. Gobi, Alash. Gobi, Ordos, Khesi.*

IIA. Junggar: *Zaisan* (between Burchum and Kaboi rivers, Kinkpai well, Karoi area, hummocky sand, June 15, 1914—Schischkin).

IIIA. Qinghai: *Amdo* ("Kao-miao bei Lotu-hsien" 1930—Hao, l.c.).

General distribution: West Sib. (south-west.), East. Sib. (south), Far East (south), Nor. Mong. (Hang.,-Mong.-Daur., Cis-Hing.), China (Dunbei, Nor., Nor.-West., Cent.).

Note 1. In the southern part of the distribution range (Alashan, Ordos), tall flexuose shoots, var. *australa* (Maxim.) Tsing et P.T. Li, are formed on sand shoals and depressions under conditions of excellent moisture.

2. *Cynanchum inamoenum* (Maxim.) Loes. has also been cited in Fl. Tibeta [Fl. Xizang, 1 (1985) 14] for Lhasa but this plant, extensively distributed in East Asia (from south. Far East to south. China), is used in Chinese medicine and is probably grown in Lhasa for these purposes.

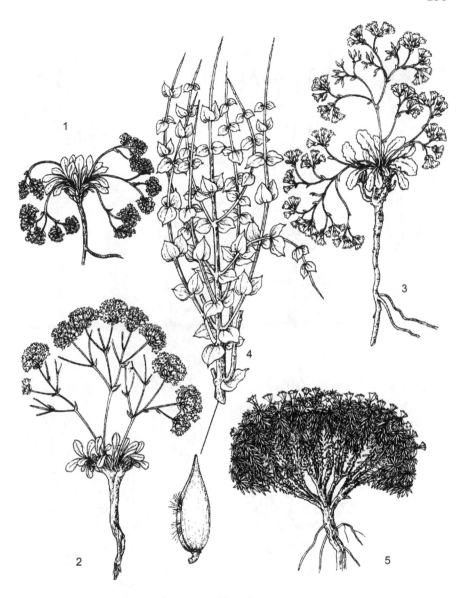

Plate I

1 — *Limonium congestum* (Ldb.) Ktze.; 2 — *L. erythrorhizum* Ik. Gal. ex Lincz.;
3 — *L. klementzii* Ik.-Gal.; 4 — *Cynanchum pusillum* Grub.;
5 — *Acantholimon hedinii* Ostenf.

134

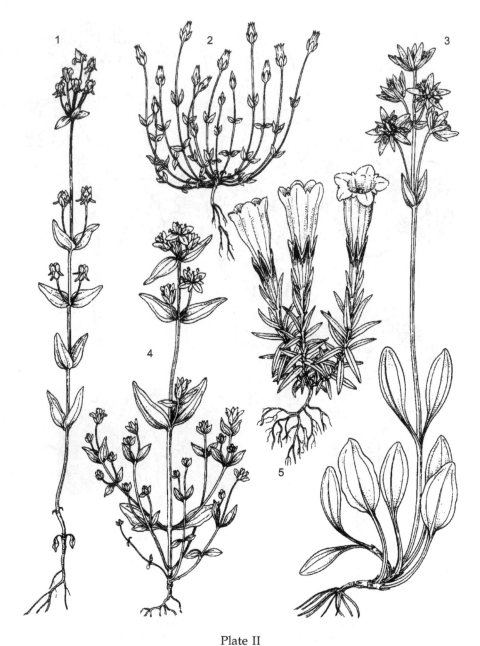

Plate II

1 — *Halenia elliptica* D. Don; 2 — *Lomatogonium brachyantherum* (Clarke) Fern.;
3 — *Swertia przewalskii* Pissjauk.; 4 — *S. tetraptera* Maxim.;
5 — *Gentiana grandiflora* Laxm.

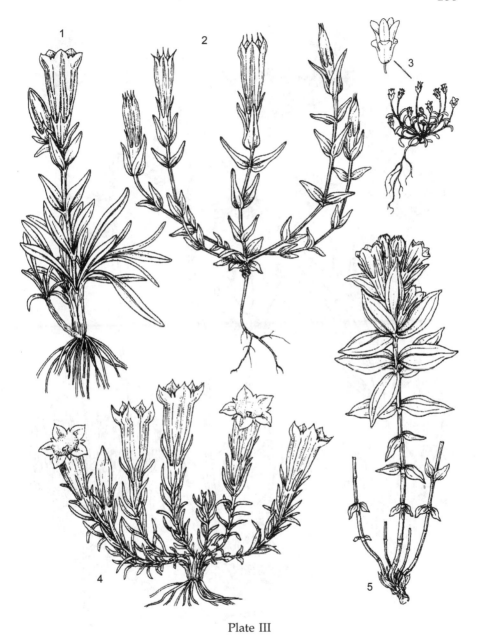

Plate III
1 — *Gentiana przewalskii* Maxim.; 2 — *G. striata* Maxim.; 3 — *G. arenaria* Maxim.;
4 — *G. farreri* Balf. f.; 5 — *G. fischeri* P. Smirn.

136

Plate IV

1— *Gentiana siphonantha* Maxim. ex Kusn.; 2— *G. aperta* Maxim.;
3— *Lomatogoniopsis alpina* Ho et Lin.; 4— *Goniolimon orthocladum* Rupr.;
5— *Ikonnikovia kaufmanniana* (Rgl.) Lincz.

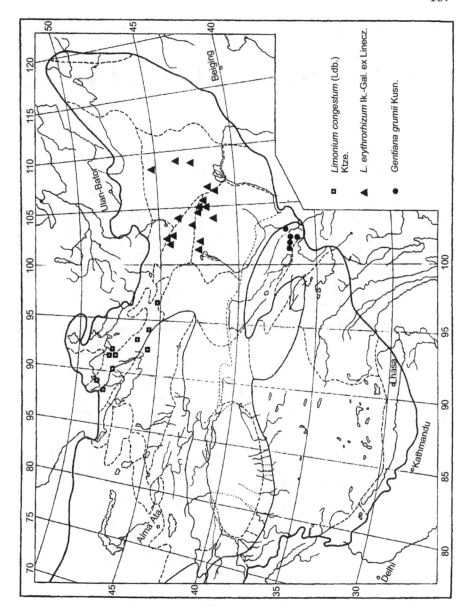

Map 1.

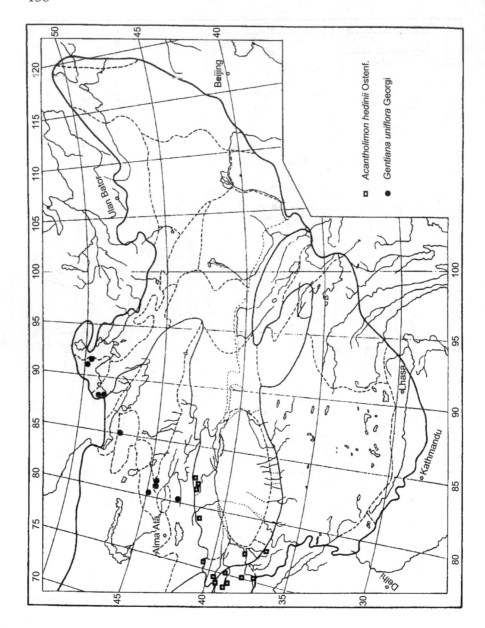

Map 2.

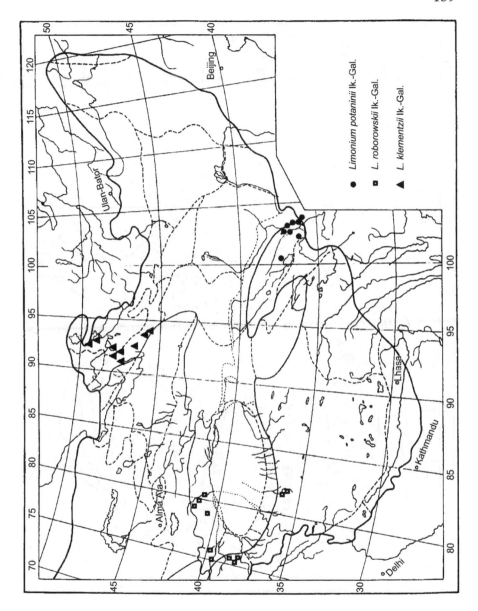

Map 3

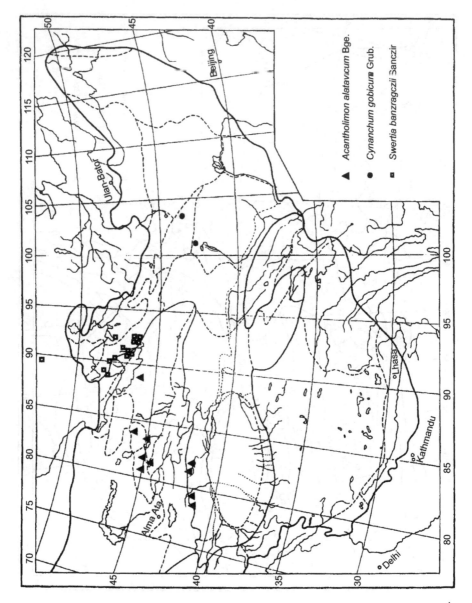

Map 4.

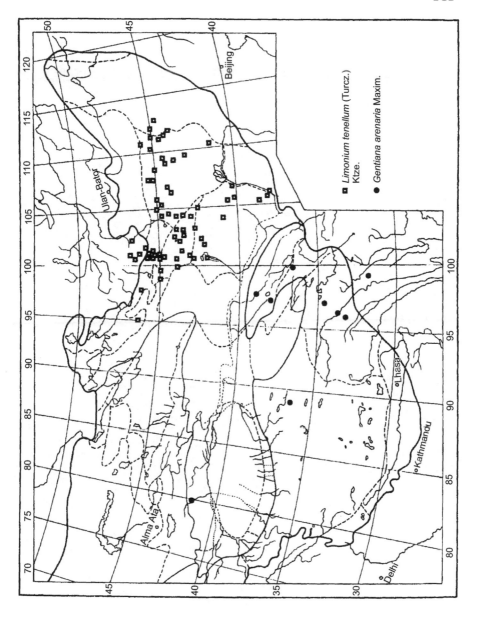

Map 5.

# INDEX OF LATIN NAMES OF PLANTS

144

146

# INDEX OF PLANT DISTRIBUTION RANGES

# INDEX OF PLANT ILLUSTRATIONS